Optimization under Constraints

Optimization under Constraints

Theory and Applications of Nonlinear Programming

PETER WHITTLE

Churchill Professor of the Mathematics of Operational Research, University of Cambridge. Fellow of Churchill College, Cambridge.

WILEY–INTERSCIENCE
a division of John Wiley & Sons Ltd
London · New York · Sydney · Toronto

Library of Congress Catalog Card Number: 75-149574

ISBN 0 471 94130 1

Printed in Great Britain by John Wright & Sons Ltd., at the Stonebridge Press, Bristol

Preface

The classic technique for dealing with constrained maximization problems is the method of Lagrangian multipliers; recent developments of this technique, together with associated numerical methods, constitute the field known as nonlinear programming. Development has been rapid in this field over the last few decades and there are many texts on the subject. These are increasingly concerned with particular numerical techniques; partly because the enormously more powerful computers of our days can accept problems in a relatively undigested form, partly because the problems of the commercial world have less structure than those of the physicist, say, and so allow less in the way of theoretical reduction. Theoretical texts tend to be pre-occupied with the case whose theory was completed earliest: linear programming. While this bias is being corrected, a theoretical text which is both inclusive and straightforward still seems lacking. In this text I have attempted to give a self-sufficient but compact account of Lagrangian theory in constrained optimization problems, taking a point of view which I hope allows both generality and simplicity. This treatment is then followed by application of the theory to a number of well-structured problems of physical, technological or economic interest. Of course, numerical methods are considered, as is both natural and necessary, but the main aim is to use the theory to derive as much detailed qualitative information as each particular case will allow, before resorting to computation.

The theory of Chapters 2 and 3 is presented in such a way as to emphasize the unity of classical Lagrangian methods with the more recent techniques of nonlinear programming. The classic results assert stationarity of the Lagrangian form at a value solving the constrained maximization problem; under appropriate convexity assumptions, such as are appealed to in linear programming, one can assert maximality of the Lagrangian form at the solution point. The two types of assertion have a geometric interpretation, which clarifies their significance and relative status. This geometric approach leads naturally to the basic concept of convexity, and to the notion of a randomized solution when convexity is lacking.

The applications covered include several which are now more or less conventional in this context: varieties of linear programme, allocation problems in both the static version and the dynamic version of a re-investment model, and flow problems, for example. Some classic topics are treated whose close relationship to the duality theory of nonlinear programming is less well-known: for instance, the deduction of Tchebichev inequalities, the magnificent Michell analysis of minimum-cost framework structures, and complementary variational principles of mathematical physics. The calculation of equilibrium distributions in statistical mechanics under restrictions on availability of energy or abundances of elements is a familiar constrained maximization problem. In Chapter 8 we achieve a generalization of Lagrangian theory by regarding such problems as limit cases of conditioned distribution problems: this work is believed to be new. Dynamic problems are not treated systematically, although a few particular problems are treated, both for their own interest and to demonstrate the relationship of the Pontryagin maximum principle to Lagrangian ideas. A systematic treatment of dynamic models is reserved for a companion volume in preparation.

Bewilderingly many numerical methods have been suggested. In this volume attention is restricted to a selected few which have a natural theoretical foundation, generality of application, and a good record in performance: the simplex method and its variants for linear problems, Fibonacci search and its variants plus the Fletcher–Powell version of the gradient method for a general unconstrained maximization problem, and sequential unconstrained methods for the constrained problem. It is probably no accident that these turn out to integrate easily with the Lagrangian theory of earlier chapters.

PETER WHITTLE

Contents

First Thoughts on Maximization

This book is concerned with the techniques for dealing with constrained maximization problems known collectively as *nonlinear programming*, the description "nonlinear" to be understood in the sense of "not necessarily linear" rather than of "not linear". However, it is as well to develop some ideas first for the case of free, or unconstrained, maximization, which is done in this opening chapter.

All the basic theory that will be used has been concentrated in the first three chapters, with only a light dilution of applications. It is thus assured that, once these chapters are understood, the rest of the book will be found lighter going.

The listing of conventions on notation has been deferred until Section (1.6), for reasons explained at the beginning of that section. However, forward reference for clarification before that point is a simple matter.

1.1 INTRODUCTION

A familiar situation is that one wishes to maximize a function $f(x)$ over some set of x-values $\mathscr{X}$, subject to the constraint

$$g(x) = b. \tag{1.1}$$

If there are several constraints, then g and b must be regarded as vectors.

The commonest example in an economic-technical context is that of *allocation*, when x represents a pattern of activity (such as the amounts of various goods which are to be manufactured), $f(x)$ the consequent economic return, $g(x)$ the consequent consumption of necessary resources (such as capital, labour, raw materials, energy, etc.) and b the amounts of these resources available. Actually, for this example one should rather modify the constraint (1.1) to

$$g(x) \leqslant b, \tag{1.2}$$

since there is usually no compulsion to exhaust all resources. This example will be extensively discussed later (see Sections (2.6), (4.3), (5.1), (6.1) and (8.1)).

Modern life and technology throw up many such problems: e.g. the optimal design of an experiment or survey, subject to cost constraints; the optimal guidance of a rocket subject to fuel constraints; the optimal design of a mechanical structure, subject to technical constraints, etc. There are also many classical examples, especially from physics. Consider, for instance, the characterization of the Gibbs distribution as the most probable distribution of energy over an assembly of molecules, subject to prescription of the values of the total number and energy of the molecules (Exercises (2.5), (8.19)).

The classical technique for dealing with such constrained extremal problems is the method of Lagrangian multipliers. This is based on the fact that, if $\bar{x}$ is a value in $\mathscr{X}$ maximizing $f(x)$ subject to (1.1), then, under some conditions, there exists a multiplier vector y with property that the form $f(x) - y^T g(x)$ is stationary at $x = \bar{x}$. This assertion, when true, we shall refer to as the *weak* or *classical Lagrangian principle*. In particular problems the multiplier y generally turns out to have a significant interpretation.

Under certain conditions a stronger conclusion can be deduced: that $\bar{x}$ maximizes the Lagrangian form $f(x) - y^T g(x)$ absolutely in $\mathscr{X}$. This assertion, when true, we shall refer to as the *strong Lagrangian principle*. The simple, global nature of the strong principle obviously makes it attractive. For instance, there is no mention of derivatives and, indeed, the principle may hold in cases when $f(x)$ or $g(x)$ do not possess derivatives at $\bar{x}$ (although compensating conditions of some other nature are required). Neither does the fact that $\bar{x}$ may be a boundary point of $\mathscr{X}$ affect the statement of the principle. Both Lagrangian principles can be adapted to the case of inequality constraints such as (1.2), or indeed, to very much more general forms of constraint (see Section (2.6)). Finally, the extremal characterization of the strong principle leads to interesting transformations of the original problem ("duality theory") and functions ("the maximum transform") in which the role of y as a variable of interest in its own right becomes apparent.

This general area is now known by the name of *nonlinear programming* ("linear" and "convex" programming being special cases). The term "programming" is an unfortunate one, easily confused with the virtually unrelated concept of computer programming, but is established. It will be used only in the first sense—that is, as referring to a maximization or optimization problem. So, when the word "programming" is used in a computational context, it will refer to numerical methods of finding a maximum or optimum. For the most part the discussion will be more concerned with the analytic aspects of programming than the computational ones. However, computational methods are too important to neglect entirely. For one thing, computation is always necessary in the last resort; for another, the best of the numerical methods are beautifully related to Lagrangian ideas. These are therefore discussed in Sections (4.5)–(4.8) and in Chapter 9.

Nevertheless, when particular applications are discussed, understanding comes before algorithms. The text therefore concentrates on analytic results and on applications which permit effective use of these: e.g. the optimization of dam regulation (Section (6.3)), of framework structures (Sections (7.2), (7.3)), of allocation and reinvestment of resources (Section (5.1)).

In Section (7.1) it will be seen how Lagrangian theory relates to the Pontryagin maximum principle of control theory, although in general consideration of dynamic problems has been reserved for another volume. An interesting further notion, scarcely to be found in the literature, is the relation of Lagrangian methods for deterministic problems to Fourier methods for stochastic problems. This is discussed, in the context of chemical equilibrium, in Sections (8.6) and (8.7).

I should have liked to have given this volume a title such as "Static Programming" or "Static Optimization", characterizing it as complementary to a projected volume on dynamic programming. However, such a title would have been pretentious, for there are so many static problems that are not touched. For instance, the complex and interesting combinatorial notions required for scheduling problems, or for optimization on a graph (e.g. routing or travelling salesman problems), are so substantial and distinct that a treatment in passing is impossible and has not been attempted. Isolated numerical methods, such as those of quadratic and integer programming, have been given no space, despite their importance. Stochastic programming is treated to some extent (Sections (6.2), (8.4) and (8.5)) but I believe that an adequate treatment of this problem would have to be in dynamic terms.

Treatment of the vector maximization problem in Chapter 10 introduces the notion of an N-person game. However, game theory is again too important and individual a subject to treat in passing, and only those aspects which relate back to Lagrangian theory have been discussed, together with a few of the interesting applications of the min–max theorem. Furthermore, game theory is another subject which can be formulated statically only up to a certain point; a satisfactory treatment would have to face the problem as a dynamic one.

The fact that so many "static" problems should properly be formulated in a dynamic model means that dynamic ideas keep breaking the surface. For this, I believe that no apology is needed.

1.2   A NOTE ON UTILITY

One sets the goal of maximizing some quantity f, without hitherto asking why. A serious attempt to answer such a question would simply raise too many issues, but a few points should be recognized.

The quality that f measures will be different in differing applications, but there are many semi-economic situations for which it has the character of an economic return, or value. f will be described as the *utility* in such cases.

The utility may be just a straight cash profit, or return. However, even in cases where money is involved, the amount of money may not be a proper measure of utility. For example, £10,000 will have more than twice the utility of £5,000 to a man who desperately needs £10,000 to avoid bankruptcy. On the other hand, the reverse may be true for a person of moderate commitments and simple tastes.

In an economic context utility is then, by definition, the quantity one would wish to maximize. It amalgamates all possible assets and measures them on a single scale. The question of the existence and proper measure of such a quantity has concerned economists and others a good deal. (Von Neumann and Morgenstern (1944) is the source reference; more accessible recent references are Chernoff and Moses (1959) and Ferguson (1967).) However, the notion will be taken for granted, except that, in Chapter 10, it is admitted that utilities may have to be vector-valued.

Additivity of utilities is sometimes assumed: that if f_1 is the utility derived from one operation and f_2 the utility from another, then the effective joint utility is just the sum, $f_1 + f_2$. This assumption may or may not be justifiable: examples of either possibility may occur to the reader.

However, a more inevitable assumption is that of *averageability*. Suppose that from an operation one may derive utility f' with probability p, or f'' with probability $q = 1 - p$. Then, it is sometimes taken as axiomatic of a utility measure that the utility of this uncertain prospect should be just the average utility, $pf' + qf''$. This condition is closely related to the notion that a utility measure should supply a relative scalar valuation of quite different prospects; the point is discussed in the works referred to above.

So, suppose that the utility f is a function $f(x)$ of x, where x describes, say, the allocation of available resources. If this allocation is uncertain and may adopt the values x' or x'' with respective probabilities p or q, then the utility of the uncertain allocation would be $pf(x') + qf(x'')$, by the principle just enunciated.

A related assumption often made is that the utility function $f(x)$ has the property

$$f(px' + qx'') \geqslant pf(x') + qf(x''), \tag{1.3}$$

i.e. that the utility of the averaged but certain allocation $px' + qx''$ be not less than that of the uncertain allocation. A function f with property (1.3) is said to be *concave*.

As yet we place no general conditions on the functions $f(x)$ with which we deal, but concavity does turn out later to be a central property.

1.3   EXTREME VALUES AND THEIR ATTAINMENT

Consider a real scalar-valued function $f(x)$, whose argument x may take values in a space whose nature need not yet be specified. Suppose that it is desired to maximize $f(x)$ freely in some x-set $\mathscr{X}$. That is, given a set of x-values $\mathscr{X}$, we wish to find a value $\bar{x}$ in $\mathscr{X}$ such that

$$f(\bar{x}) \geqslant f(x) \qquad (1.4)$$

for all x in $\mathscr{X}$. If the goal is minimization then we look for the reverse inequality: a minimizing value $\underline{x}$ such that

$$f(\underline{x}) \leqslant f(x). \qquad (1.5)$$

Now, if $\mathscr{X}$ is a finite set, then the extremizing values $\bar{x}$ and $\underline{x}$ will certainly exist. However, they need not exist if $\mathscr{X}$ is infinite. Consider, for example, the case of a man who wishes to catch a train by the least possible margin. Let $f(x)$ be his waiting time if he arrives a time x before the train is due to depart. Attention will be restricted to $x \geqslant 0$, so that $\mathscr{X}$ is the positive real line, and $f(x) = x$ for $x > 0$. Suppose that he misses the train if he arrives at the exact departure time, and must then wait a unit time interval for the next train. The full prescription of $f(x)$ in $\mathscr{X}$ is then

$$f(x) = \begin{cases} x & (x > 0), \\ 1 & (x = 0). \end{cases} \qquad (1.6)$$

Now, such a function has no attainable minimum. The minimum is certainly not reached at $x = 0$, but for any positive x there is always another value (say $x/2$) for which f is smaller. There is thus no $\underline{x}$ satisfying (1.5) for all $x \geqslant 0$.

The trouble arises because the formulation makes a distinction that could never be realized in the physical world, a distinction between x zero and x positive but arbitrarily small. Furthermore, the distinction is important: it represents the difference between missing or catching the train.

Such artificial distinctions and sensitivities are the usual reason for failure of an extremum to exist, and they can usually be eliminated by a more careful and realistic formulation of the problem (see Exercise (1.2)).

For a somewhat similar example, consider the case of a man who has cornered the world resources of uranium, of amount a, say, and releases these on to the market at rate x. Then, $\mathcal{X}$ is again the non-negative real line. If the price realized in this way is $p(x)$, then the total value of sales is

$$f(x) = \begin{cases} 0 & (x = 0), \\ ap(x) & (x > 0). \end{cases} \tag{1.7}$$

The owner wishes to choose x so as to maximize this quantity. Now, $p(x)$ will be non-negative and may reasonably be supposed decreasing in x, since commodities are more expensive when scarce. If $p(x)$ is strictly decreasing for small positive x, then it follows as before that $f(x)$ has no maximum in $\mathcal{X}$.

Again, the trouble arises from the artificiality of the formulation. More realistic analysis would show that there are several factors which exclude the notion of a vanishingly small rate of sale: the cost to the owner of maintaining a sales organization, his desire to realize his assets in a finite time, the fact that the market might have no use for a commodity available at less than a certain rate (or might simply not accept the situation!).

In cases where a maximum or minimum does not exist, one introduces the more general idea of a *supremum* or *infimum*. The supremum of $f(x)$ in $\mathcal{X}$, denoted $\sup f(x)$, is the least value of λ for which

$$f(x) \leqslant \lambda \tag{1.8}$$

for all x in $\mathcal{X}$. Correspondingly, the infimum, denoted $\inf f(x)$, is the greatest value of λ for which

$$f(x) \geqslant \lambda \tag{1.9}$$

for all x in $\mathcal{X}$.

So, for the train-catching example, $\inf f(x) = 0$. For the uranium-monopoly

$$\sup f(x) = a \lim_{x \downarrow 0} p(x);$$

the limit existing in virtue of the monotone nature of $p(x)$. These values represent extremes that can be approached arbitrarily closely, but not attained.

The statement "the maximum exists" is equivalent to the statement "the supremum is attained": correspondingly for the minimum and infimum.

Sufficient conditions for a function to attain its extreme values are given in the following theorem.

Theorem (1.1)

If $f(x)$ is continuous, and $\mathcal{X}$ a compact set in a finite-dimensional Euclidean space, then $f(x)$ attains its extreme values in $\mathcal{X}$.

The result will only be proved for $\mathcal{X}$ a finite closed interval on the real line: the interested reader can either generalize this proof or refer to Apostol (1957, p. 73) for discussion of the more general case.

Denote the interval by I_0, and let the supremum of $f(x)$ over the interval be $\bar{f}$. Divide the interval into two equal closed sub-intervals (so that the division point belongs to both). The supremum of f on one or other of these sub-intervals must equal $\bar{f}$. Denote the sub-interval for which this is so by I_1 (making an arbitrary choice if the supremum equals $\bar{f}$ for both sub-intervals). Continuing in this way, a sequence $\{I_j\}$ is generated of nested, closed intervals with lengths tending to zero, f having supremum $\bar{f}$ on every I_j. If x_j is an arbitrarily chosen point in I_j, then the sequence $\{x_j\}$ is plainly convergent. Its limit point $\bar{x}$ must belong to I_0 (i.e. to $\mathscr{X}$), since I_0 is closed.

Now,

$$f(\bar{x}) \leqslant \bar{f}, \tag{1.10}$$

by the definition of $\bar{f}$. On the other hand, since $\bar{f}$ is also the supremum of f on I_j, there is a value of x in I_j for which $f(x)$ exceeds the value $\bar{f} - \varepsilon$, where ε is an arbitrary positive quantity. We can choose x_j so that it is just such a value, so that then

$$f(x_j) \geqslant \bar{f} - \varepsilon. \tag{1.11}$$

Since f is continuous we can assert that

$$|f(x_j) - f(\bar{x})| \leqslant \delta_j \quad (j = 1, 2, \ldots), \tag{1.12}$$

where $\{\delta_j\}$ is a sequence tending to zero. From (1.11), (1.12) it follows that

$$f(\bar{x}) \geqslant \bar{f} - \varepsilon - \delta_j \quad (j = 1, 2, \ldots). \tag{1.13}$$

But ε may be chosen arbitrarily small, and $\{\delta_j\}$ tends to zero, so relations (1.10), (1.13) imply that $f(\bar{x}) = \bar{f}$: the supremum $\bar{f}$ is attained at the point $\bar{x}$ of $\mathscr{X}$. The theorem is thus proved.

Exercises

(1.1) Show that $f(x)$ has a maximum if $\mathscr{X}$ is finite.

(1.2) Consider the train-catching example discussed in the text. Suppose that the time when the train becomes no longer boardable is random, and distributed about the advertized departure time with a known continuous probability density and a finite mean value. Let x be the time elapsing between arrival of the passenger and cessation of boarding, and let us suppose that

$$f(x) = \begin{cases} x & (x > 0), \\ 1 & (x \leqslant 0). \end{cases}$$

Show that there is a choice of arrival time which minimizes the expected value of $f(x)$. (Appeal to Theorem (1.1).)

(1.3) Consider a function $f(x, y)$ of two variables taking values in sets $\mathscr{X}, \mathscr{Y}$ respectively. Suppose that $f(x, y)$ attains its maximum in $\mathscr{Y}$ for any fixed

x in $\mathcal{X}$, that $\max_y f(x,y)$ attains its maximum in $\mathcal{X}$, and correspondingly for x, y reversed. Show then that

$$\max_x \max_y f(x,y) = \max_y \max_x f(x,y).$$

1.4   THE LOCATION OF EXTREME VALUES

In most cases it will be supposed that the function f is such that it possesses a maximum in $\mathcal{X}$ (possibly attained at several points of $\mathcal{X}$). The maximum value will be denoted by $U = f(\bar{x})$. The taking of a maximum (of f over $\mathcal{X}$) can be regarded as an operation on the function f, which will be written as

$$U = \max f(x), \tag{1.14}$$

or $\max_x f(x)$ if the relevant set is in doubt. (See Section (1.6) for a summary of conventions on notation.)

Correspondingly, one has the minimizing operation, $\min f(x)$. Because of the identity

$$\min f(x) = -\max\left[-f(x)\right] \tag{1.15}$$

a minimization problem can always be rephrased as one of maximization. In general discussions maximization can thus be taken as being the standard extremal problem.

Now, conceivably, a maximizing value $\bar{x}$ could be located by testing all the points of $\mathcal{X}$ in turn for the maximizing property (1.4): the *enumerative* method of maximization. This blind method of search is applicable only for finite $\mathcal{X}$, and even then it may be prohibitively time-consuming (see Exercise (1.4)). However, one may be able to exploit special properties of f to show right from the beginning that only certain points of $\mathcal{X}$ could possibly yield a maximum, so that Theorem (1.1) can be restricted to members of a sub-set of $\mathcal{X}$. Such special properties may be analytic ones, such as continuity or differentiability (see the next section), or they may be structural ones specific to the example.

As an example of appeal to structural properties, consider a variant of the so-called "travelling-salesman" problem. The salesman wishes to visit each of M given points in the plane, starting from and returning to a particular one of these points, in such a way as to minimize the total distance traversed. It will be assumed there is no restriction on the paths in the plane, but, in order to avoid degeneracies, that no three of the points to be visited are collinear.

One wishes then to minimize $f(x)$, where x labels a possible path, and $f(x)$ is the length of this path. The set $\mathcal{X}$ of possible paths is very large, but simple arguments quickly show that attention can be confined to a relatively small sub-set of it.

To begin with, attention can be restricted to routes which follow straight lines between points. Secondly, those routes can be excluded which visit any point more than once. The number of possibilities has thus been reduced to the $(M-1)!$ simple circuits of the M points, with straight-line connections. In fact, most of these can be eliminated; for a route which intersects itself cannot be optimal (Exercise (1.5)).

The problem is not yet solved, but it has been enormously reduced by these simple arguments. Further such arguments lead to an efficient computational solution of the problem. (For a discussion of such problems, see Part II of Berge and Ghouila-Houri (1965).)

Exercises

(1.4) Suppose that the set $\mathscr{X}$ considered is finite, with N members. Show that exactly $N-1$ comparisons of pairs of f values are needed, to find the maximum of f in $\mathscr{X}$ by the enumerative method.

(1.5) Demonstrate the validity of the reductions of the travelling-salesman problem in the text. For the final point: if two arcs intersect, forming the diagonals of a quadrilateral Q, consider using two opposite sides of Q instead of the diagonals as elements of the path.

(1.6) The reductions of the travelling-salesman problem all followed from the assertion of a relation $f(Tx) \leqslant f(x)$ for some operator T. What were the operators invoked?

1.5 STATIONARITY

Suppose that $\mathscr{X}$ is a set in a Euclidean space of n dimensions, so that x can be represented as a column vector of n elements: $(x_1, x_2, ..., x_n)$. A partial test of whether a particular value $\bar{x}$ truly maximizes $f(x)$ in $\mathscr{X}$ can be made by considering *local variations*, i.e. by applying the test (1.4) for x in the neighbourhood of the postulated $\bar{x}$.

Suppose that the point $\bar{x}+\varepsilon s$, where ε is a non-negative scalar and s a vector, belongs to $\mathscr{X}$ for all non-negative ε less than some positive value $\varepsilon(s)$. Then s will be described as *directed into the interior of $\mathscr{X}$ from $\bar{x}$*, or as a *feasible direction* from $\bar{x}$. Denote the row vector of derivatives

$$\left[\frac{\partial f}{\partial x_1}, \frac{\partial f}{\partial x_2}, ..., \frac{\partial f}{\partial x_n}\right],$$

if this exists, by f_x.

Theorem (1.2)

If the derivative f_x exists at $\bar{x}$, then

$$f_x s \leqslant 0 \qquad\qquad (1.16)$$

at $\bar{x}$ for feasible directions s from $\bar{x}$. In particular, if $\bar{x}$ is an interior point of $\mathscr{X}$, then

$$f_x = 0 \qquad\qquad (1.17)$$

at $\bar{x}$.

Criterion (1.17) is the classic *stationarity condition*, the condition that $\bar{x}$ be a *stationary point* of f. Its use in extremal problems goes back to Kepler and Fermat. The criterion is an invaluable one, even though there are many important problems whose solution does not lie at a stationary point.

To prove the theorem, we appeal to the partial Taylor expansion

$$f(\bar{x} + \varepsilon s) = f(\bar{x}) + \varepsilon f_x s + o(\varepsilon), \qquad\qquad (1.18)$$

If ε is sufficiently small, then $\bar{x} + \varepsilon s$ will lie in $\mathscr{X}$, and we can assert that

$$f(\bar{x} + \varepsilon s) \leqslant f(\bar{x}). \qquad\qquad (1.19)$$

Relations (1.18), (1.19) imply that $f_x s + o(1) \leqslant 0$, whence inequality (1.16) follows in the limit $\varepsilon \downarrow 0$.

If $\bar{x}$ is an interior point of $\mathscr{X}$, then (1.16) holds for any s. Setting s equal to v and $-v$ in this inequality, we deduce that

$$f_x v = 0 \qquad\qquad (1.20)$$

for any v, whence (1.17) follows.

The conclusion of the theorem can be expressed in the somewhat weakened form:

Corollary

A maximizing point $\bar{x}$ of f is either (i) a stationary point of f, (ii) a point where f is non-differentiable or (iii) a boundary point of $\mathscr{X}$.

So, the class of conceivable maximizing points has been reduced to this extent. Note that possibility (iii) does not exclude (i) or (ii).

The exercises provide several examples of use of the stationarity criterion. In the meanwhile, note that all three cases of the Corollary can present themselves quite easily in practice. For example, suppose that, if a firm decides to produce goods at rate x, then the value of its sales will amount to $a \log(x+1)$, and costs will be $bx + c[x]$, where $[x]$ is the smallest integer not less than x. The $\log(x+1)$ term represents decreasing (less than linear) returns; the $[x]$ term represents discontinuities caused by the capital cost of plant which must be installed in units of a minimum size. It is desired now to choose a non-negative value of x that will maximize net profit:

$$f(x) = a\log(x+1) - bx - c[x].$$

Depending on the relative values of the coefficients a, b and c, the return f may reach its maximum value at the boundary point $x = 0$ (if no positive level of activity is profitable), at a stationary point (when one unit of plant is in partial use) or at an integral value of x, at which f is non-differentiable (when all units are in full production, but to add another would be uneconomic). See Figure (1.1).

That f has a stationary point at $\bar{x}$ does not imply that it is maximal at $\bar{x}$, or even that it has a local maximum there (i.e. that (1.4) holds for x in some neighbourhood of $\bar{x}$). For example, the stationarity condition (1.17) could equally be consistent with the existence of a local minimum at $\bar{x}$. Again, the function

$$f(x_1, x_2) = x_1^2 - x_2^2$$

is stationary at the origin, but increases with $|x_1|$ and decreases with $|x_2|$, so that the origin is neither a maximum nor a minimum, but a *saddle-point*.

Stationarity is thus a necessary condition for a maximum if $\bar{x}$ is a value in the interior of $\mathscr{X}$ at which f is differentiable, but it is by no means sufficient. However, the previous argument can be continued to deduce some rather stronger conditions.

Theorem (1.3)

Suppose f maximal at a point $\bar{x}$ interior to $\mathscr{X}$. If the matrix of second derivatives f_{xx} exists at $\bar{x}$, then, in addition to condition (1.17),

$$s^T f_{xx} s \leqslant 0 \tag{1.21}$$

for any vector s. That is, $-f_{xx}$ is non-negative definite at $\bar{x}$.

The assertion follows by extension of the expansion (1.18) to

$$f(\bar{x} + \varepsilon s) = f(\bar{x}) + \varepsilon f_x s + \tfrac{1}{2}\varepsilon^2 s^T f_{xx} s + o(\varepsilon^2), \tag{1.22}$$

and repetition of the previous argument.

Exercises

(1.7) Let x_k be the labour force of a factory in month k: it is required to change this from an initial value x_0 to a final value L by month n as quickly and yet as smoothly as possible. To this end, it is decided to minimize

$$\sum_1^n [(x_k - x_{k-1})^2 + \lambda(x_k - L)^2]$$

with respect to $x_1, x_2, ..., x_{n-1}$. Show that the solution $\bar{x}$ satisfies

$$\bar{x}_{k+1} - 2\bar{x}_k + \bar{x}_{k-1} = \lambda(\bar{x}_k - L) \quad (k = 1, 2, ..., n-1).$$

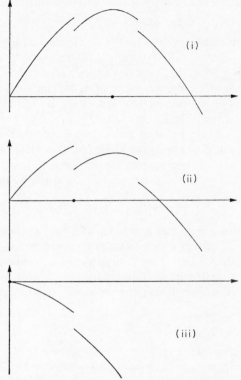

Figure 1.1 An illustration of the different possible characters of a maximizing point: (i) maximum attained at a stationary point (for the example of the text, one unit of plant is in partial use); (ii) maximum attained at a point where the function has no derivative (all units are in full production, but to add another would be uneconomic); (iii) maximum attained at a boundary point of the permitted set (no positive level of production is profitable)

This difference equation is to be solved subject to the end-conditions $\bar{x}_0 = x_0$, $\bar{x}_n = L$.

(1.8) An alternative formulation of requirements in Exercise (1.7) might be: minimize $\sum_1^n (x_k - L)^2$ subject to $|x_k - x_{k-1}| \leqslant A$, and x_0 prescribed. What is the solution? Under what conditions is $x_n = L$ attained?

(1.9) Let $\mathscr{X}$ be the positive orthant, $x \geqslant 0$, in n-dimensional Euclidean space, and suppose that f is everywhere differentiable. Show that at the maximizing point $\bar{x}$, $\partial f / \partial x_k \leqslant 0$, with equality if $\bar{x}_k > 0$ ($k = 1, 2, ..., n$).

(1.10) *Horse-backing.* Consider a racecourse in which a proportion α of the money staked is distributed to bettors, it being divided among backers of the winning horse in proportion to the size of bets placed. Before an n-horse race a bettor knows that a total stake t_k has been placed on horse k, which has a probability p_k of winning ($k = 1, 2, ..., n$). Show that, if he places a bet x_k on this horse ($k = 1, 2, ..., n$), then his expected net gain is

$$f(x) = \alpha \left(\sum \frac{p_k x_k}{x_k + t_k} \right) [(\sum (x_k + t_k)] - \sum x_k.$$

He wishes to choose x (in $x \geqslant 0$) to maximize this. Show that there exists a constant c such that

$$\bar{x}_k = t_k \left[c \sqrt{\left(\frac{p_k}{t_k} \right)} - 1 \right]_+$$

where ξ_+ denotes the positive part of ξ (see equation (1.26)). Find a determining relation for c. Show that the bettor never backs all horses.

(1.11) Consider an electrical network, consisting of a number of nodes j, with nodes j and k connected by a direct link of resistance R_{jk}. If y_j is the electrical potential at node j, then the rate of dissipation of energy in the network is proportional to

$$D = \tfrac{1}{2} \sum \sum \frac{(y_j - y_k)^2}{R_{jk}}.$$

Certain of these potentials will be prescribed: the others then assume the values that minimize D. Show that if y_j is not prescribed, then

$$\sum_k (y_j - y_k) / R_{jk} = 0. \tag{1.23}$$

What form does this equation take if the network takes the form of a plane rectangular grid, with connections (each of unit resistance) to nearest neighbours only, and potentials prescribed only along the edge of the grid?

(1.12) Consider the least-square approximation of a function $\phi(\omega)$ by a linear combination $\sum_1^n \alpha_k \psi_k(\omega)$ of functions $\psi_k(\omega)$. That is, the coefficients α_k are chosen so as to minimize the mean-square difference,

$$\Delta(\alpha) = \int [\phi(\omega) - \sum \alpha_k \psi_k(\omega)]^2 \, W(\omega) \, d\omega,$$

where W is an appropriate weighting function, and all integrals are assumed to exist. The aim is thus to minimize $\Delta(\alpha)$, which can be written as a quadratic form in its arguments:

$$\Delta(\alpha) = A - 2B^\mathrm{T} \alpha + \alpha^\mathrm{T} C \alpha$$

Show that the matrix C is non-negative definite. Hence show that any solution $\bar{\alpha}$ of the stationarity equation

$$C\alpha = B$$

satisfies $\Delta(\bar{\alpha}) \leqslant \Delta(\alpha)$. In this case, then, stationarity is both a necessary and sufficient condition for a minimizing value.

(1.13) Suppose that $\bar{x}$ is on the boundary of $\mathscr{X}$, that f is differentiable at $\bar{x}$, and that the class of inward-directed vectors at $\bar{x}$ includes a linear manifold $\mathscr{M}$. (That is, if s and s' belong to $\mathscr{M}$, then so does $\lambda s + \mu s'$, for any pair of scalars λ, μ.) Show that $f_x s = 0$ for s in $\mathscr{M}$.

(1.14) Continuing Exercise (1.13), show that, if f possesses second derivatives at $\bar{x}$, then $s^T f_{xx} s \leqslant 0$ for s in $\mathscr{M}$.

(1.15) Suppose that inequality (1.21) is strict for non-zero s, so that $-f_{xx}$ is positive definite. Show that f has then at least a local maximum at $\bar{x}$.

(1.16) A special case of Exercise (1.12) is that in which α is chosen so as to minimize $\int [\phi(\omega) - \alpha]^2 W(\omega) d\omega$, when we find $\bar{\alpha} = \int \phi W d\omega / \int W d\omega$, the *arithmetic mean* of ϕ with weighting W. As an example of a case where derivatives may not exist, consider the choice of α to minimize $\int |\phi(\omega) - \alpha| W(\omega) d\omega$. Note that under certain conditions all α-values in some interval may solve the problem.

(1.17) Consider the continuous version of Exercise (1.7): to minimize

$$I(x) = \int_0^T \left[\left(\frac{dx}{dt}\right)^2 + \lambda(x-L)^2 \right] dt$$

with respect to the function $x(t)$, subject to $x(0)$ and $x(T)$ prescribed. By assuming a twice-differentiable solution, and considering alternatives of the form $\bar{x}(t) + \varepsilon s(t)$, where $s(t)$ is once-differentiable, show that

$$\frac{d^2 \bar{x}(t)}{dt^2} = \lambda(\bar{x}(t) - L) \quad (0 < t < T)$$

(Derive an expression

$$I(\bar{x} + \varepsilon s) = I(\bar{x}) + \varepsilon \int A(t) s(t) dt + o(\varepsilon)$$

where A does not depend on s.) In fact, this equation determines the true minimizing function.

(1.18) Continuous versions of the network problem of Exercise (1.11) can be similarly considered. If $y(\xi, \eta)$ is the electrical potential at Cartesian co-ordinate (ξ, η) on a uniformly conducting plane sheet, then y minimizes

$$\iint \left[\left(\frac{\partial y}{\partial \xi}\right)^2 + \left(\frac{\partial y}{\partial \eta}\right)^2 \right] d\xi \, d\eta,$$

subject to prescription of value at certain points. Suppose that y is prescribed on the boundary of a region C and free inside C. Show then, by a formal variational argument of the type used in Exercise (1.17), that one might expect

$$\frac{\partial^2 y}{\partial \xi^2} + \frac{\partial^2 y}{\partial \eta^2} = 0$$

within C (*Dirichlet's principle*). In fact, this equation does determine the true minimizing function if the boundary of C and the y-values prescribed upon it are sufficiently smooth. This they may not always be, although it could be again argued that the cases when Dirichlet's principle fails are just those in which the mathematics draws distinctions which are non-realizable physically.

1.6 CONVENTIONS ON NOTATION, ETC.

This section would logically have been placed at the beginning. However, being neither easily digestible nor part of the main course, it has been deferred to this point, where the reader will already have made a start on the true subject matter, and will have picked up some of the conventions incidentally.

Neither matrices nor vectors are distinguished by bold type. An n-vector x, with elements $(x_1, x_2, ..., x_n)$ will be understood to be a column-vector. Its transpose, the corresponding row-vector, will be denoted x^{T}. The transpose of a matrix A is correspondingly denoted A^{T}. The kth element of the column vector Ax will be denoted $(Ax)_k$.

By $x \geqslant 0$ it is meant that all elements of x are non-negative:

$$x_k \geqslant 0 \quad (k = 1, 2, ..., n). \tag{1.24}$$

By $x > 0$ it is meant that (1.24) holds, with strict inequality for at least one value of k. By $x \gg 0$ it is meant that strict inequality holds throughout:

$$x_k > 0 \quad (k = 1, 2, ..., n). \tag{1.25}$$

Correspondingly for $\leqslant$, $<$, etc. Occasionally the negations of these signs will be used, e.g. $\not>$, with a consistent interpretation.

An n-dimensional Euclidean space will be denoted R^n. The positive orthant of this space (i.e. those points with co-ordinate $x \geqslant 0$) will be denoted R_+^n.

If $f(x)$ is a scalar function of an n-vector variable x, then f_x will denote the *row* vector of partial derivatives $\partial f / \partial x_k$, and f_{xx} the square matrix of second derivatives $\partial^2 f / \partial x_j \, \partial x_k$. If $g(x)$ is an m-vector function of an n-vector argument, then g_x will denote the $m \times n$ matrix of derivatives $\partial g_j(x) / \partial x_k$. The $m \times n$ arrangement is consistent with the convention that g_x is a row vector in the case $m = 1$.

The statement "$x \in \mathcal{X}$" is to be read "x belongs to the set $\mathcal{X}$", and "$x \notin \mathcal{X}$" is the negation of this statement. If $\mathcal{X}, \mathcal{C}$ are sets, then $(\mathcal{X}, \mathcal{C})$ denotes the compound set of elements (x, z) with $x \in \mathcal{X}$ and $z \in \mathcal{C}$.

The conventional set notation is employed. If $\mathcal{A}, \mathcal{B}$ are sets in a common space (i.e. sub-sets of a given set) then the intersection $\mathcal{A} \cap \mathcal{B}$ is the set of points lying in both, the union $\mathcal{A} \cup \mathcal{B}$ is the set of points lying in either, and the inclusion relationship $\mathcal{A} \subset \mathcal{B}$ states that all elements of $\mathcal{A}$ belong to $\mathcal{B}$. If addition of elements is defined, then $\mathcal{A} + \mathcal{B}$ is the set of elements $\xi + \xi'$, with $\xi \in \mathcal{A}$ and $\xi' \in \mathcal{B}$.

If $f(x)$ is a function of x, and x adopts values in a set $\mathcal{X}$, then $f(\mathcal{X})$ will denote the set with elements $f(x)$ for some x in $\mathcal{X}$. Correspondingly, if g is a function and $\mathcal{X}, \mathcal{C}$ are sets, then $g(\mathcal{X}) + \mathcal{C}$ will denote the set of elements $g(x) + z$ for some x in $\mathcal{X}$ and some z in $\mathcal{C}$. (It is thus necessary that $g(x)$ and z take values in the same space, say R^m, with addition defined.)

If x is a scalar, then x_+ denotes the "non-negative part of x":

$$x_+ = \begin{cases} x & (x \geqslant 0), \\ 0 & (x \leqslant 0). \end{cases} \tag{1.26}$$

If x is an n-vector with elements x_k, then x_+ denotes the n-vector with elements $(x_k)_+$.

The operations of taking a maximum, minimum, supremum or infinum will be denoted max, min, sup and inf respectively. For explicitness the variable with respect to which the operation is being performed should be indicated, and also its domain. For example, $\max_{x \in \mathcal{X}} L(x, y)$, the maximum of $L(x, y)$ as x is varied over the set $\mathcal{X}$. However, mention is sometimes omitted of either the variable, its domain, or both, if the context is such that this will cause no confusion. So, the last case might be written $\max_x L(x, y)$ if the point is that the maximum is to be taken with respect to x, its domain of variation being understood. Correspondingly, we might write $\max_x f(x)$ if it is clear that the maximization is with respect to x, and all that needs emphasis is that x is being varied over the set $\mathcal{X}$.

The pair of letters p, q are invariably employed to denote a pair of non-negative real numbers adding to unity. More generally, (p_k) denotes a *distribution*, a set of non-negative real numbers p_k adding to unity.

CHAPTER 2

Constrained Maximization and Lagrangian Methods

The classic technique for dealing with constrained maximization problems is the method of Lagrange multipliers. We give a treatment of this method which leads naturally on to more recent developments of the technique.

2.1 CONSTRAINED MAXIMIZATION; LAGRANGIAN MULTIPLIERS

Consider now the problem which is to form the principal study: the maximization of a function $f(x)$, subject to $x \in \mathscr{X}$, as before, but also to the functional constraints

$$g(x) = b. \qquad (2.1)$$

Here g and b may be vectors (of m elements, say) if there are several constraints.

In many cases the constraint (2.1) expresses the fact that the optimization is being carried out subject to limited resources. The important special problem of optimal allocation has already been mentioned in Section (1.1), and it is useful to take this as the typical problem when one wishes to think in concrete terms. A standard formulation of the problem is that different activities may be undertaken (e.g. the manufacture of different types of article, or the channelling of capital into different types of investment) and it is desired to distribute effort over these activities in such a way as to maximize financial return. This maximization is constrained by the fact that the activities should not jointly consume more resources (e.g. raw materials, time, labour, energy) than are available. Let x_k denote the rate at which the kth activity is pursued, and x, the *intensity vector*, be the vector of these n quantities. Since rates are inherently non-negative, x is restricted to the

17

non-negative orthant $x \geqslant 0$, which constitutes the basic set $\mathscr{X}$ for this problem. $f(x)$ represents the return from an allocation x, the element $g_j(x)$ of $g(x)$ represents the amount of resource j consumed by such an allocation, and the element b_j of b represents the amount of this resource available. The object is then to choose x in R_+^n to maximize $f(x)$, subject to condition (2.1).

Actually, since a compulsion to use all resources will usually neither exist nor be economically desirable, for this example the constraint (2.1) should be modified to the inequality constraint

$$g(x) \leqslant b. \tag{2.2}$$

For instance, a brewery which had access to limited amounts of grain and hops, but to a good supply of water, would not do well if it felt obliged to incorporate all this water in its products.

It will be seen in Section (2.6) that methods for dealing with the equality constraint (2.1) generalize naturally to cases with more general types of constraint, such as (2.2). However, for the moment only the case of equality constraints will be considered.

There is a fundamental and powerful method for dealing with such constrained problems: the method of Lagrangian multipliers. The method depends upon the fact that, under certain conditions, a vector of multipliers $y = (y_1, y_2, \ldots, y_m)$ will exist such that the Lagrangian form $f(x) - y^T g(x)$ is stationary with respect to x at the solution $\bar{x}$ of the constrained maximization problem. An appropriate stationary point of the Lagrangian form is located, and y is then adjusted until the constraint (2.1) is satisfied at the stationary point.

The standard elementary justification of the method goes somewhat as follows. Consider a variation $\bar{x} \to \bar{x} + \varepsilon s$ of the solution point in a feasible direction s, which is compatible with (2.1). Assuming g differentiable, then in the limit of small ε,

$$g_x s = 0. \tag{2.3}$$

Since f is not increased by this variation then

$$f_x s \leqslant 0, \tag{2.4}$$

as in Theorem (1.2).

Condition (2.3) is plainly a necessary condition on s, the direction of perturbation of solution. Suppose that it is also sufficient, in that any s satisfying (2.3) is a possible direction of perturbation. This assumption implies that any such s is feasible, and also implies a local regularity condition which will be discussed later (see Exercise (2.1) and Section (2.4)).

Now, if s satisfies (2.3), so does $-s$. It follows then, as in Theorem (1.2), that (2.4) may be strengthened to

$$f_x s = 0 \qquad (2.5)$$

for any s satisfying (2.3). This can be true only if f_x is linearly dependent on the row vectors of g_x, i.e. if

$$f_x - y^T g_x = 0 \qquad (2.6)$$

for some vector y. That is, there is a vector y for which the Lagrangian form $f - y^T g$ is stationary at $\bar{x}$.

The proof is a not unenlightening one, but it is not surprising that people who have known no other should find the Lagrangian multiplier a somewhat mysterious notion to the end of their days. In fact, the multiplier has a very clear significance, both mathematically and physically. Further, these ideas can be strengthened (for example: $\bar{x}$ may actually *maximize* the Lagrangian form in $\mathscr{X}$). To obtain this deeper understanding, a rather different setting of the problem and type of proof is needed, which will be prepared in the next two sections.

The reader may well fail to see the point of introducing multipliers, when there are two obvious methods of dealing with the constrained maximization problem. One is to simply regard the permissible region $\mathscr{X}$ as being reduced to a smaller one, $\mathscr{X}(b)$ say, by the specification of the additional constraints (2.1). Another is to use constraint (2.1) to express some of the components of x in terms of the others, and so eliminate them, and achieve a free maximization problem in fewer variables.

These measures do work occasionally, but there are many cases where an attempt at such a direct reduction of the problem leads either to impracticable calculations or to a severe loss of symmetry. In general, the cases where it is natural to cope with the constraint (2.1) by use of Lagrangian multipliers are those in which it is natural to imbed the problem in a family of constrained maximization problems, by considering the constraint (2.1) for variable b. For example, for the allocation problem b is the vector of amounts of resources available, and there might well be an interest in solving the problem for a range of values of b, and of regarding both the optimal allocation $\bar{x}$ and the maximal return as functions of b. This point of view will be illuminated in the following sections.

Exercises

(2.1) Prove that condition (2.3) is sufficient (i.e. any s satisfying (2.3) is a possible direction of perturbation of solution) if $\bar{x}$ is interior to $\mathscr{X}$ and $g(x)$ is linear.

That some local regularity condition on g may be necessary can be seen from the example with a single constraint

$$\sum_1^n x_k{}^2 = 0.$$

Plainly $\bar{x} = 0$, since $x = 0$ is the only point satisfying the constraint. At this point $g_x = 0$, so the local linear approximation (2.3) to the condition fails to constrain s at all. This point is discussed further in Section (2.4).

Use the multiplier technique to solve the problems in Exercise (2.2)–(2.5), justifying its application.

(2.2) Minimize $\sum_1^n v_k/x_k$ in $x \geqslant 0$, subject to $\sum a_k x_k = b$. The constants a_k, v_k and b are all strictly positive. (This example concerns the minimization of the variance of an estimate derived from a stratified sample survey subject to a cost constraint. x_k is the size of sample for the kth stratum, and a_k, v_k are measures of sampling cost and of variability for this stratum.)

(2.3) Minimize the positive definite quadratic form $c^{\mathrm{T}} V c$ with respect to the vector c, subject to the constraint $c^{\mathrm{T}} c = 1$. Show that the minimal value, the Lagrangian multiplier and the minimal eigenvalue of the matrix V are all equal. (This problem arises from an attempt to fit a linear relationship $\sum c_k \zeta_k \sim 0$ to a number of random variables $\zeta_1, \zeta_2, ..., \zeta_n$. The coefficients c_k are chosen, subject to the normalization condition mentioned, so as to minimize the average value of $(\sum c_k \zeta_k)^2$, which equals $c^{\mathrm{T}} V c$ if V_{jk} is the average value of $\zeta_j \zeta_k$.)

(2.4) Minimize $c^{\mathrm{T}} V c$ with respect to c subject to the constraint $\xi^{\mathrm{T}} c = 1$, where ξ is a given vector. (This problem occurs in the statistical estimation of a postulated linear relationship, disturbed by random error, between two quantities. See the chapter on linear models in Lindgren, 1960.)

(2.5) Consider the maximization of

$$f(x) = \sum_{k=1}^n (x_k - x_k \log x_k)$$

in $x \geqslant 0$, subject to

$$\sum x_k = \mathrm{N},$$

$$\sum x_k \varepsilon_k = \mathscr{E},$$

where N, $\mathscr{E}$ and the ε_k are strictly positive. (The motivation for this problem is the interpretation of x_k as the number of molecules in a certain volume possessing energy ε_k, the total number N and total energy $\mathscr{E}$ being prescribed. The probability of a partition x over energy levels is proportional to $\prod_k (x_k!)^{-1}$ in the unconstrained case, if the molecules do not interact; $f(x)$

is an approximation (for large x_k) to the logarithm of this quantity. Thus $\bar{x}_k$, the *Gibbs distribution*, describes the most probable distribution of molecule numbers over energy levels in the limit of large N. This example is considered further in Sections (3.1), (3.7) and (8.7).)

(2.6) Note that a maximization subject to (2.2) and $x \in \mathscr{X}$ can be rephrased a maximization with respect to x, z subject to $g(x) + z = b$, and $x \in \mathscr{X}$, $z \geqslant 0$. That is, the inequality constraint can be rewritten as an equality constraint by the introduction of the *slack variable z*. This is a useful device, as will be seen in Section (2.6). However, the somewhat different status of the two sets of variables x and z must be observed if the structure of the problem is not to be obscured.

2.2 FAMILIES OF CONSTRAINED PROBLEMS; GEOMETRIC CONCEPTS

The constrained maximization problem of the last section is to be imbedded, then, in a family of such problems. Let the problem of maximizing $f(x)$ in $\mathscr{X}$ subject to $g(x) = b$ be denoted by $P(b)$. Let a maximizing value, hitherto written $\bar{x}$, now be denoted $x(b)$, and let the maximum value attained, $f(x(b))$, be denoted $U(b)$.

The problem and its solution are thus being regarded as functions of b. For the allocation problem, this means looking at the solution as a function of resources available; for the sample survey problem of Exercise (2.2), as a function of available finance; for the energy distribution problem of Exercise (2.5), as a function of average molecule density and energy (equivalently, density and temperature).

The x-values in $\mathscr{X}$ which satisfy $g(x) = b$ will be termed *feasible* values for $P(b)$, and the set of such values will be denoted $\mathscr{X}(b)$. These are the values which satisfy the basic conditions of the problem, and which are candidates for the solution of $P(b)$. (For feasible *directions*, see note on p. 39).

If a solution is to exist (i.e. if $\mathscr{X}(b)$ is not to be empty) then there must exist at least one feasible x-value. This means that the only set of b-values that should be considered are those in the set

$$\mathscr{B} = g(\mathscr{X}), \tag{2.7}$$

the set generated by $g(x)$ as x takes values in $\mathscr{X}$. For example, there would be no point in prescribing negative b-values in an allocation problem, for which $g(x)$ is inherently positive. Values of b in $\mathscr{B}$ will be termed *attainable*.

The function $U(b)$ thus has $\mathscr{B}$ as natural domain of definition. The properties of this function are crucial: its characterizing property is that given in Exercise (2.7).

An object of even greater interest than $\mathscr{B}$ is the set $\mathscr{D} = (g(\mathscr{X}), f(\mathscr{X}))$ in R^{m+1}. That is, the set in a Euclidean space with co-ordinates

$$(\xi_1, \xi_2, \ldots, \xi_m, \eta) = (\xi, \eta)$$

whose points have co-ordinates

$$\left. \begin{array}{l} \xi = g(x), \\ \eta = f(x) \end{array} \right\} \tag{2.8}$$

for some x in $\mathscr{X}$. For the allocation problem, for example, $\mathscr{D}$ is the set of values of resource consumption $g(x)$ and corresponding economic return $f(x)$ which are technically possible; which can be achieved by some allocation x in the range of possibilities $\mathscr{X}$. However, in studying $\mathscr{D}$ alone one suppresses the x-dependence; the means by which a particular value of $(\xi, \eta) = (g, f)$ is attained. All that is retained is knowledge of the set of (g, f) values which is attainable. However, this set is informative, and its geometry generates the Lagrangian techniques.

Special interest attaches, of course, to the "upper" boundary of $\mathscr{D}$; those points (ξ, η) which belong to $\mathscr{D}$ but which are such that $(\xi, \eta + \delta)$ does not belong to $\mathscr{D}$ for any positive δ. These are the optimal points; the points for which f is maximal for an assigned value of g. They satisfy the equation

$$\eta = U(\xi) \tag{2.9}$$

which is the graph of the upper boundary of $\mathscr{D}$.

In order to continue, the notions of a supporting hyperplane and a tangent hyperplane to a set at a prescribed boundary point are needed. Consider a point-set $\mathscr{A}$ in a Euclidean space, and a point Q on its boundary. A hyperplane in the space is said to be a *supporting hyperplane* to $\mathscr{A}$ at the boundary point Q if it passes through Q, but lies wholly to one side of $\mathscr{A}$, as in cases (i), (iv) of Figure (2.1). (A more exact definition will be given in Section (3.2).) A hyperplane is a *tangent hyperplane* to $\mathscr{A}$ at Q if the distance between any other boundary point Q' and the hyperplane is of smaller order than the distance $Q' - Q$ for $Q' - Q$ small, as in cases (i), (ii) of the figure. A tangent hyperplane is evidently *locally supporting* in the sense that points of $\mathscr{A}$ within a small radius d of Q lie to one side of the hyperplane, to within amounts of magnitude $o(d)$.

A hyperplane may be both tangent and supporting (case (i)), tangent without being supporting, (case (ii)), supporting without being tangent (case (iv)), or $\mathscr{A}$ may be such that neither type exists at a given boundary point (case (iii)).

What we shall show in the next section is that, broadly speaking, validity of the strong (weak) Lagrangian principle introduced in Section (1.1) corresponds to the existence of a supporting (tangent) hyperplane to $\mathscr{D}$ at the appropriate boundary point. The multiplier vector y is just the gradient of this hyperplane relative to the η-axis in the (ξ, η) co-ordinate system. If

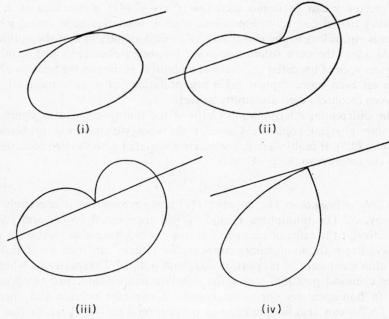

(i)   (ii)

(iii)   (iv)

Figure 2.1 A boundary point of a set may possess a hyperplane which is both supporting and tangent (case i), tangent without being supporting (case ii), supporting without being tangent (case iv), or no hyperplane of either character may exist (case iii)

the hyperplane is tangent to $\mathscr{D}$ then this will also be the gradient of the graph of $U(\xi)$ at b:

$$y_j = \frac{\partial U}{\partial b_j} \qquad (2.10)$$

(see Exercise (2.8)). This identification gives one some feeling for the significance of the multipliers, and gives them their "price" interpretation in economic contexts. For the allocation problem $\partial U / \partial b_j$ would be the rate of change of financial return with change in amount of resource j available, an optimal allocation being preserved under all circumstances. It thus represents a fair marginal price for resource j.

2

Of course, a price interpretation emerges already (in the allocation case) when the problem of maximizing $f(x)$ constrainedly is modified to that of maximizing $f(x) - y^T g(x)$ unconstrainedly in $x \geqslant 0$; this latter quantity would be optimizer's net profit if he could buy resources freely at prices y_j. If a vector y can be found such that $f(x) - y^T g(x)$ is maximal at $\bar{x}$, the solution to the constrained problem, then a price has been found whose effect is equivalent to the constraints of limited supply, in that the optimizer would adopt the same solution vector $\bar{x}$ for the cases of unlimited resources at prices y, or of uncosted resources of amount b. However, the relation (2.10) gives an even more explicit price interpretation: of y as a trade-off rate between resources used and utility gained.

The differential interpretation (2.10) of the multipliers is also significant in more physical contexts. Consider the energy-distribution problem of Exercise (2.5). If multipliers y_1 and y_2 are associated with the two constraints, then the solution to the problem is

$$\bar{x}_k = e^{-y_1 - y_2 \varepsilon_k}, \tag{2.11}$$

the *Gibbs distribution*. The quantity $f(x)$ being maximized is effectively the entropy, S. The multipliers y_1 and y_2 will then equal $\partial S / \partial N$ and $\partial S / \partial \mathscr{E}$ respectively: the rates of change of entropy with changes in total mass and energy. From thermodynamic considerations these can then be identified, to within a constant of proportionality, with μ and $1/T$ respectively, where μ is the chemical potential and T the absolute temperature. For the present case (a homogeneous one-phase system at constant volume and energy) $y_1 = \partial S / \partial N$ can also be identified as proportional to $-F/T$, where F is the free energy.

If U_b does not exist, so that the hyperplane is supporting without being tangent, as in case (iv) of Figure (2.1), the simple identification (2.10) is no longer possible. However, an identification and a price interpretation persist (see Section (3.6)): what happens is that prices may depend upon the particular direction in which b is changed.

Exercises
(2.7) Show that

$$U(g(x)) \geqslant f(x)$$

for any x in $\mathscr{X}$, with equality for some x for every attainable value of the argument of U.

(2.8) Show from Exercise (2.7) that, if the derivatives f_x, g_x and U_b exist at $x(b)$ and b respectively, then

$$(f_x - U_b g_x) s \leqslant 0$$

for all feasible directions s from $x(b)$. Thus a version of the Lagrangian principle in some respects more general than that discussed in Section (2.1) is obtained, and also, the identification $y^T = U_b$.

(2.9) Show that for Exercise (2.2) $\mathscr{B}$ is the positive real axis, and that, if $L(b)$ is the minimal value of the criterion function as a function of b, then

$$L(b) = [\sum (a_k v_k)^{\frac{1}{2}}]^2/b.$$

What is $\mathscr{D}$?

(2.10) Consider the maximization of x_1 in the x_1/x_2 plane subject to $x_1^2 + x_2^2 = b$. Then $\mathscr{B} = R_+^1$; show that $U(b) = \sqrt{b}$.

(2.11) Consider the maximization of x_1 in the plane set $x_2 \geqslant x_1^2$ subject to $x_2 = b$. Then again, $\mathscr{B} = R_+^1$ and $U(b) = \sqrt{b}$.

(Exercises (2.10) and (2.11) may appear too simple to have much point: their point is their illustration that $U(b)$ may have infinite derivative at boundary points of $\mathscr{B}$, which later proves to be important.)

(2.12) Consider the maximization of

$$f(x) = \alpha x_1 + x_2^{\beta+1}/(\beta+1)$$

in $x_1, x_2 \geqslant 0$, subject to $x_1 + x_2 = b$. The constants α and $\beta+1$ are strictly positive. Determine the maximizing point, and show that

$$U(b) = \begin{cases} b^{\beta+1}/(\beta+1), & (b \leqslant \alpha^{1/\beta}), \\ \alpha b - \dfrac{\beta}{\beta+1}\alpha^{(\beta+1)/\beta}, & (b \geqslant \alpha^{1/\beta}) \end{cases}$$

if $\beta < 0$, but that

$$U(b) = \max(\alpha b, b^{\beta+1}/(\beta+1))$$

if $\beta \geqslant 0$.

2.3   A GEOMETRICAL CHARACTERIZATION OF THE LAGRANGIAN PRINCIPLES

The classical assertion, established under certain conditions in Section (2.1), is: there exists a vector y such that $f(x) - y^T g(x)$ is stationary in x at the solution point $x(b)$ of $P(b)$. This might be called the *classic* or *weak* Lagrangian principle, in contrast to the *strong* principle, which, when valid, asserts that $f(x) - y^T g(x)$ is maximal in $\mathscr{X}$ at $x = x(b)$.

In order to include certain extreme cases it is useful to modify these principles slightly, as follows.

The homogeneous weak Lagrangian principle: There exists a vector y and a non-negative scalar w such that inequality (2.19) holds at $x(b)$, the solution to the constrained optimization problem, for all feasible directions s from

$x(b)$. (See the note on p. 39). In particular, if $x(b)$ is interior to $\mathcal{X}$, then $wf(x) - y^T g(x)$ is stationary at $x(b)$.

The homogeneous strong Lagrangian principle: There exists a vector y and a non-negative scalar w such that $wf(x) - y^T g(x)$ is maximal for x in $\mathcal{X}$ at $x(b)$.

The validity of these principles and the interpretation of the multipliers can be related to the properties of the set $\mathcal{D} = (g(\mathcal{X}), f(\mathcal{X}))$, and, in particular, of its upper boundary, described by (2.9). For typographical simplicity the solution to $P(b)$ will be denoted by $\bar{x}$ rather than $x(b)$. The following results are asserted for a fixed value of b.

Theorem (2.1)   (see Figure (2.2))

(i) *The homogeneous strong Lagrangian principle is valid for $P(b)$ if and only if a supporting hyperplane to $\mathcal{D}$ exists at the point $(b, U(b))$.*

(ii) *The coefficient w may be normalized to unity if b is interior to $\mathcal{B}$, and $U(\xi)$ is finite in a neighbourhood of b.*

(iii) *If b is interior to $\mathcal{B}$ and the vector of derivatives U_b exists at b, then, with the normalization $w = 1$, it is true that*

$$y^T = U_b. \tag{2.12}$$

The first part of the theorem is trivial, because assertion of the existence of a supporting hyperplane is just a thinly disguised statement of the strong

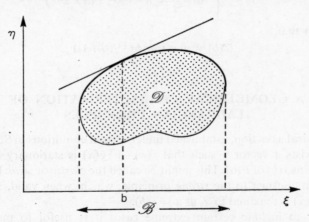

Figure 2.2   An illustration of the set $\mathcal{D} = (g(\mathcal{X}), f(\mathcal{X})$. The "upper" boundary of the set (with graph $\eta = U(\xi)$) is heavily drawn. The case has been represented in which $\mathcal{D}$ possesses a supporting hyperplane at the point $(b, U(b))$

Lagrangian principle. It is the last conclusion which is significant: the identification of y_j with $\partial U/\partial b_j$, if this exists.

If the strong principle is valid, then w, y exist such that

$$wf(\bar{x}) - y^{\mathrm{T}} g(\bar{x}) \geqslant wf(x) - y^{\mathrm{T}} g(x) \quad (x \in \mathscr{X}) \tag{2.13}$$

or

$$wU(b) - y^{\mathrm{T}} b \geqslant w\eta - y^{\mathrm{T}} \xi \quad ((\xi, \eta) \in \mathscr{D}). \tag{2.14}$$

But relation (2.14) just states that the hyperplane in (ξ, η) space

$$w(\eta - U(b)) - y^{\mathrm{T}}(\xi - b) = 0, \tag{2.15}$$

which passes through the boundary point $(b, U(b))$ of $\mathscr{D}$, lies to one side of $\mathscr{D}$. That is, it is a supporting hyperplane to $\mathscr{D}$ at $(b, U(b))$.

Conversely, if a supporting hyperplane (2.15) exists it will be characterized by an inequality (2.14), which then implies the Lagrangian characterization (2.13). Choice of the direction of inequality in (2.14) as $\geqslant$ rather than $\leqslant$ corresponds to a convention $w \geqslant 0$. For, if $\delta > 0$ then the point

$$(\xi, \eta) = (b, U(b) - \delta)$$

lies on the same side of the hyperplane as does $\mathscr{D}$. It therefore lies in the region (2.14), whence $w\delta \geqslant 0$, or $w \geqslant 0$.

If it can be established that $w > 0$, then w can be normalized to unity, as in the initial formulation of the strong Lagrangian principle. The case $w = 0$ is that in which the hyperplane (2.15) is vertical, i.e. parallel to the η-axis. If $\mathscr{D}$ is bounded this can occur only at the "edge" of the upper boundary of $\mathscr{D}$, i.e. only if b is a boundary point of $\mathscr{B}$. Formally: suppose that $w = 0$. If both $U(\xi)$ and $U(b)$ are finite, then it follows from (2.14) that

$$y^{\mathrm{T}}(\xi - b) \geqslant 0. \tag{2.16}$$

Under the assumptions of part (ii) of the theorem, (2.16) may be asserted for all ξ in some neighbourhood of b, whence it follows that $y = 0$. But, if the hyperplane is to exist, as stipulated, w and y cannot both be zero. The case $w = 0$ is thus excluded, and w can be normalized to unity.

In particular, we can achieve the normalization if b is interior to $\mathscr{B}$ and the derivative U_b exists at b. Setting $w = 1$ and

$$\eta = U(\xi) = U(b) + U_b(\xi - b) + o(|\xi - b|) \tag{2.17}$$

in inequality (2.14) we deduce that

$$(y^{\mathrm{T}} - U_b)(\xi - b) + o(|\xi - b|) \geqslant 0. \tag{2.18}$$

Since this holds for all ξ in some neighbourhood of b, the final conclusion (2.12) follows.

Theorem (2.2)

(i) *Suppose a tangent hyperplane to $\mathscr{D}$ exists at $(b, U(b))$, and that the derivatives f_x, g_x exist at $\bar{x}$. Then a vector y and a non-negative scalar w exist such that*

$$(wf_x - y^{\mathrm{T}} g_x)s \leqslant 0 \tag{2.19}$$

for all feasible directions s from $\bar{x}$, so the homogeneous weak Lagrangian principle is valid.

(ii) *If U_b exists then one can normalize w to unity, and y^{T} then necessarily equals U_b.*

A hyperplane tangent to $\mathscr{D}$ at $(b, U(b))$ is assumed to exist; let this be denoted by (2.15), with the convention $w \geqslant 0$. The locally supporting property of a tangent hyperplane then implies that (2.14) holds in the modified form

$$wU(b) - y^{\mathrm{T}} b \geqslant w\eta - y^{\mathrm{T}} \xi + o(|\xi - b, \eta - U(b)|) \tag{2.20}$$

for (ξ, η) in $\mathscr{D}$. Here $|\xi, \eta|$ denotes the length of the vector (ξ, η) in R^{m+1}. In particular,

$$wf(\bar{x}) - y^{\mathrm{T}} g(\bar{x}) \geqslant wf(x) - y^{\mathrm{T}} g(x) + o(|g(x) - g(\bar{x}), f(x) - f(\bar{x})|) \tag{2.21}$$

for x in $\mathscr{X}$. Because of the differentiability assumption, the remainder term in (2.21) can be written $o(|x - \bar{x}|)$. Setting $x = \bar{x} + \varepsilon s$ and letting ε tend to zero, we then deduce the inequality (2.19). Part (ii) of the theorem follows as in Theorem (2.1).

No converse statement is made in Theorem (2.2), to the effect that validity of the weak Lagrangian principle would imply the existence of a tangent hyperplane to $\mathscr{D}$. Such an assertion can only be made if further assumptions are introduced, ensuring that the solution $x(b)$ is itself differentiable in b, or something similar.

Still, the broad relationship exists between the Lagrangian principles and the surface geometry of $\mathscr{D}$ (or, equivalently, the nature of the function $U(b)$). What is now needed is to translate the conditions of Theorems (2.1) and (2.2) on $\mathscr{D}$, which are not immediately verifiable, into conditions on $f(x)$, $g(x)$ and $\mathscr{X}$ which are. Some results will be obtained in this direction for the weak principle in the next section, and for the strong principle in the next chapter.

Exercises (2.10) and (2.11) provide examples of cases where U_b becomes infinite as b tends towards the boundary of $\mathscr{B}$. The boundary case ($b = 0$, in both examples) would usually be taken as an instance in which Lagrangian methods fail, in that the multiplier y is necessarily infinite. However, in the homogeneous form the principles still hold (with zero w instead of infinite y).

Exercises

(2.13) Consider the maximization of $x_1{}^2 - x_2{}^2$ in the x_1/x_2 plane subject to $x_1 = b$. What is the form of $U(b)$? Show that the Lagrangian principle is valid in the weak form, but not in the strong.

(2.14) Consider the example of Exercise (2.12). Note that the cases $\beta \leqslant 0$, $\beta > 0$ correspond respectively to $\mathscr{D}$s having the form (i), (iii) of Figure (2.1). Thus, both Lagrangian principles are valid in the first case, but only the weak in the second, and even that not valid at $b = [\alpha(\beta + 1)]^{1/\beta}$ (at which point there is a discontinuity in the solution, $x(b)$).

2.4 SUFFICIENT CONDITIONS FOR VALIDITY OF THE WEAK LAGRANGIAN PRINCIPLE

The characterization of the Lagrangian principles in terms of the geometric properties of the set $\mathscr{D}$ (more particularly of the graph of the function $U(b)$) is undoubtedly illuminating and fundamental. However, as it stands the characterization yields less information than it does insight, in that it appeals to properties of the derived quantity $U(b)$ rather than of the given quantities $f(x)$, $g(x)$ and $\mathscr{X}$. If sufficient conditions for the validity of the Lagrangian principles are to be derived, the geometrical approach of Section (2.3) must be supplemented by arguments relating the properties of $U(b)$ to those of $f(x)$, $g(x)$ and $\mathscr{X}$. The nature of these supplementary arguments may depend considerably on special features of the case considered. This is particularly true for the weak principle, considered in this section. The strong principle admits a more coherent treatment, given in Chapter 3.

Theorem (2.3)

The homogeneous weak Lagrangian principle is valid if the maximizing point $x(b)$ is interior to $\mathscr{X}$, and the derivatives f_x, g_x exist at this point. The coefficient w may be normalized to unity if either of the following conditions is fulfilled in addition: (i) g is linear, or (ii) g_x is of full rank m.

It is already known from the argument of Section (2.1) that there exists a y such that $f - y^{\mathrm{T}} g$ is stationary at $x(b)$ if equation (2.3) is a sufficient local expression of the constraints $g(x) = b$ and $x \in \mathscr{X}$, i.e. if all directions s satisfying condition (2.3) represent a possible direction of perturbation $x(b) \to x(b) + \varepsilon s$ of the solution. This will certainly be true if $x(b)$ is interior to $\mathscr{X}$ and $g(x)$ is linear, so the theorem is established in case (i).

In case (ii) the functions $g_1, g_2, \ldots, g_m$ can be supplemented by $n - m$ others to form a vector $\tilde{g} = (g_1, g_2, \ldots, g_n)$ such that the transformation $x \to \tilde{g}$ has a non-vanishing Jacobian. There will then be a non-degenerate neighbourhood of $x(b)$ which maps on to a non-degenerate neighbourhood of $\tilde{g}(x(b))$ in

$\tilde{g}$-space. If $f(x)$ is maximal under local x-variations, subject to $g(x) = b$ it must also be maximal under the corresponding local $\tilde{g}$ variations, i.e. under free local variation of $g_{m+1}, g_{m+2}, \ldots, g_n$. Now, by the usual chain rule

$$\frac{\partial f}{\partial x_k} = \sum_{j=1}^{n} \frac{\partial f}{\partial g_j} \frac{\partial g_j}{\partial x_k}. \tag{2.22}$$

But, by what has just been said, $\partial f/\partial g_j = 0$ for $j > m$. Relation (2.22) thus implies stationarity of the form $f - y^{\mathrm{T}} g$, with the identification $y_j = \partial f/\partial g_j$.

All that remains now is to prove the first part of the theorem, for which neither of the conditions (i) or (ii) is invoked. If condition (ii) is not fulfilled then it fails, i.e. there exists a vector y such that

$$y^{\mathrm{T}} g_x = 0 \tag{2.23}$$

In this case the homogeneous weak principle holds trivially, with $w = 0$.

The restriction that the solution be interior to $\mathscr{X}$ is an embarrassment, for the solution may well lie on the boundary of $\mathscr{X}$ in many cases. Moreover, Theorem (2.2) gives a lead as to how the weak principle generalizes in such a case, the stationarity condition being replaced by a first-order maximality condition (2.19) for all feasible directions s from $x(b)$.

It can be seen from elementary treatment of a simple case that such a generalization must hold in certain cases. Consider the maximization of $f(x)$ in $x \geqslant 0$ subject to

$$\sum_{1}^{n} x_k = b, \tag{2.24}$$

a one-constraint allocation problem. Suppose f differentiable in all arguments, and, for simplicity, denote the solution as previously by $\bar{x}$.

If $b > 0$ then $\bar{x}_j > 0$ for at least one j. Consider the perturbation, consistent with (2.24), $\bar{x}_j \to \bar{x}_j + \delta$, $\bar{x}_k \to \bar{x}_k - \delta$ for particular values j, k, all other components of $\bar{x}$ remaining unchanged. Since f cannot increase under a feasible perturbation we have

$$\delta\left(\frac{\partial f}{\partial x_j} - \frac{\partial f}{\partial x_k}\right) \leqslant 0, \tag{2.25}$$

the differentials being evaluated at $x = \bar{x}$. If $\bar{x}_j$ and $\bar{x}_k$ are both positive then δ can have either sign, and it follows from (2.25) that $\partial f/\partial x_j = \partial f/\partial x_k$. That is, $\partial f/\partial x_j$ has a common value, say y, for all j such that $\bar{x}_j > 0$.

If $\bar{x}_j = 0$ and $\bar{x}_k > 0$ then δ can only be chosen positive: it follows then from (2.25) that

$$\frac{\partial f}{\partial x_j} \leqslant y. \tag{2.26}$$

So, at the optimal point inequality (2.26) holds for all j, with equality if $\bar{x}_j > 0$. But this is just the statement that

$$(f_x - y1)s \leqslant 0 \qquad (2.27)$$

for any feasible direction of perturbation s of the solution point, where 1 is row-vector of units. The Lagrangian form thus certainly obeys the condition (2.19) in this particular case.

This example worked out easily from first principles. Nevertheless, it is not a completely straightforward matter to prove that (2.19) is a necessary condition for optimality in a moderately general case. The nub of the proof of Theorem (2.3) is the fact, appealed to in Section (2.1), that if $f_x s \leqslant 0$ for all s such that $g_x s = 0$, then f_x is linearly dependent on the rows of g_x. This fact is easily established, but the generalization of it which is needed requires new ideas for its proof.

Lemma

Suppose that $f_x s \leqslant 0$ for all s such that $g_x s = 0$ and $s \in \mathcal{K}$, where $\mathcal{K}$ is a convex cone. Then there exists a vector y and a non-negative scalar w such that $(wf_x - y^T g_x)s \leqslant 0$ for all s in $\mathcal{K}$. The constant w can be normalized to unity if (i) 0 is interior to $g_x \mathcal{K}$, or (ii) $\mathcal{K}$ is polyhedral.

A convex cone is a set closed under positive linear operations, in that, if s and s' both belong to it, then so does $\lambda s + \mu s$, where λ, μ are non-negative scalars. A polyhedral cone is a set specified by a finite number of linear inequalities of the type $\beta^T s \geqslant 0$. The lemma is a version of Farkas' lemma; we shall defer its proof until Section (3.3). Appealing to it, we obtain the following generalization of Theorem (2.3) immediately.

Theorem (2.4)

Suppose that the derivatives f_x, g_x exist at the maximizing point $x(b)$, and that the set of feasible directions s from $x(b)$ forms a convex cone $\mathcal{K}$. Then there exists a vector y and a non-negative scalar w such that

$$(wf_x - y^T g_x)s \leqslant 0 \qquad (2.28)$$

for all s in $\mathcal{K}$. The constant w can be normalized to unity if either of the conditions (i), (ii) of Theorem (2.3) holds and either of the conditions (i), (ii) of the lemma holds.

Exercises

(2.15) Consider the maximization of $f(x)$ in the rectangular interval $(0 \leqslant x_k \leqslant M_k; k = 1, 2, ..., n)$ subject to condition (2.24). Show from first principles that a y exists satisfying (2.27) for all feasible directions of perturbation s of the optimal solution.

(2.16) Show that, for the multiplier y associated with the constraint (2.24)

$$y\delta \leqslant U(b+\delta) - U(b) + o(\delta)$$

and hence that $y = U_b$, if this exists. The value of y can thus be regarded as the marginal rate of return from an investment b. This characterization casts a revealing light on condition (2.26). If the jth activity is not undertaken ($\bar{x}_j = 0$) then the rate of return $\partial f / \partial x_j$ on this activity will certainly not exceed the overall rate y: if the activity is undertaken ($\bar{x}_j > 0$) then $\partial f / \partial x_j$ precisely equals the overall rate.

(2.17) Appealing to Theorem (2.4), obtain the analogues of conditions (2.26), (2.27) and of the assertions of Exercise (2.16) for the multi-constraint allocation problem: maximization of $f(x)$ in $x \geqslant 0$ subject to $Ax = b$.

(2.18) Condition (2.28) is a first-order maximality condition on the Lagrangian form $f(x) - y^T g(x)$. It may be asked whether the form can always be asserted to be at least locally maximal in x at $x(b)$, when y has the appropriate value. Show from the example of Exercise (2.13) that this is not necessarily true. Show more particularly that the form is locally maximal for variations consistent with $g(x) = b$, but not necessarily for others.

(2.19) Suppose that $x(b)$ is interior to $\mathcal{X}$, and that all necessary derivatives of $f(x)$, $g(x)$, $U(b)$ and $x(b)$ exist. Let H denote the matrix of second derivatives $(f - y^T g)_{xx}$, and U_{bb} denote the corresponding matrix for U. Show that $U_{bb} = x_b^T H x_b$ and $H x_b = g_x^T U_{bb}$, whence

$$U_{bb} = (g_x H^{-1} g_x^T)^{-1}$$

if these matrix inverses exist. Show further that the matrix

$$H^* = H - g_x^T (g_x H^{-1} g_x^T)^{-1} g_x$$

satisfies $H^* x_b = 0$, and is non-positive definite. This is another way of making the last assertion of Exercise (2.18).

2.5 AN APPLICATION: CROSS-CURRENT EXTRACTION

As an example, consider the optimization of a continuous chemical process by Lagrangian methods. The example is interesting as a physically well-motivated problem for which the Lagrangian principle is valid in the weak form, although not generally in the strong.

The process is that of cross-current extraction, in which solute is extracted from a stream of solvent by repeated washings with water. The solvent stream is passed consecutively through a sequence of extractors, in each of which a cross-current of wash-water, flowing at a determined rate, carries out some of the solute. The aim is to choose the individual wash-flow rates in such a way as to extract as much solute as possible by the end, the total rate of flow of wash-water being prescribed (see Figure (2.3)).

So, let there be n extractors, labelled $k = 1, 2, ..., n$, and let x_k be the concentration of solute in the solvent stream leaving extractor k. The initial

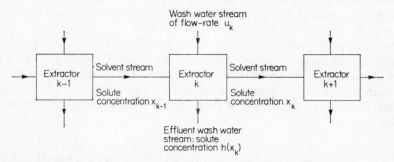

Figure 2.3 A diagram of the sequence of extractors, and of the solvent and wash-water streams

value x_0 is given, and it is desired to minimize the terminal concentration x_n. Let Q be the solvent flow-rate, and u_k the flow-rate of wash-water in extractor k. One supposes that the concentrations of solute in solvent and wash-streams leaving extractor k are in chemical equilibrium, so that the concentration of solute in the wash-stream is a known function of x_k, say $h(x_k)$. Then the solvent balance equation for extractor k is

$$Q(x_{k-1} - x_k) - u_k h(x_k) = 0 \quad (k = 1, 2, ..., n). \tag{2.29}$$

There is also a limitation on total wash-water rate:

$$\sum_{1}^{n} u_k = W, \tag{2.30}$$

say. (Strictly, an inequality restriction, $\sum u_k \leqslant W$, but the problem is such that all wash-water available is in fact used.)

The problem can then be posed as the minimization of $f(x, u) = x_n$ with respect to x_k, u_k $(k = 1, 2, ..., n)$, subject to the constraints (2.29) and (2.30). If these two sets of n elements are denoted by the vectors x, u, then there is a natural limitation to the domain $x \geqslant 0$, $u \geqslant 0$. The nature of the function h is obviously important. For the moment it will be merely supposed that $h(0) = 0$, and that $h'(x)$ exists and is strictly positive for $0 \leqslant x < \infty$. If the left-hand member of (2.29) is denoted by g_k, then the transformation

$$(x, u) \rightarrow (g_1, g_2, ..., g_n, \sum u_k, u_2, u_3, ..., u_n)$$

is readily seen to possess a non-singular Jacobian, so that the condition (ii) of Theorem (2.3) is fulfilled, as far as constraints (2.29) and (2.30) are concerned. Stationarity of a Lagrangian form at the solution point $(\bar{x}, \bar{u})$ can thus be asserted if it can be shown that this solution point is interior to the region $(x, u) \geqslant 0$. That is, that $\bar{x}_k > 0$, $\bar{u}_k > 0$ $(k = 1, 2, ..., n)$.

It is seen from (2.29) that x_k is zero only if either x_{k-1} is zero or u_k is infinite: strict positivity of the x_k is thus assured.

Zero u_k values would correspond to the fact that some extractors were not being used. In such a case, one could assume without loss of generality that $u_1 = 0, u_2 > 0$: i.e. extractor 1 is not in use, but extractor 2 is. Consider now an infinitesimal redistribution of wash-water over the first two extractor stages: $(u_1, u_2) \rightarrow (u_1 + \varepsilon, u_2 - \varepsilon)$. It then follows from the implicit relationship (2.29) that

$$\frac{dx_2}{d\varepsilon} = -\frac{\partial x_2}{\partial u_2} + \frac{\partial x_2}{\partial x_1}\frac{\partial x_1}{\partial u_1} = \frac{1}{Q + u_2 h'(x_2)}\left[h(x_2) - \frac{Qh(x_1)}{Q + u_1 h'(x_1)}\right].$$

Then at $u_1 = \varepsilon = 0$

$$\frac{dx_2}{d\varepsilon} = \frac{h(x_2) - h(x_1)}{Q + u_2 h'(x_2)} < 0,$$

so that the reallocation is advantageous. That is, $u_1 = 0$ cannot be optimal, and all extractors will be used in the optimal solution.

Appeal to the weak Lagrangian principle is thus legitimate, and one can assert stationarity of the form

$$x_n - \sum_1^n y_k[Q(x_{k-1} - x_k) - u_k h(x_k)] - v\sum_1^n u_k \qquad (2.31)$$

at the solution point. The multipliers y_k and v account for constraints (2.29) and (2.30) respectively. The stationarity conditions are

$$u_k y_k h'(x_k) = Q(y_{k+1} - y_k) \quad (k = 1, 2, ..., n-1), \qquad (2.32)$$

$$u_n y_n h'(x_n) = -Qy_n - 1, \qquad (2.33)$$

$$y_k h(x_k) = v \quad (k = 1, 2, ..., n). \qquad (2.34)$$

Relations (2.29), (2.32) and (2.34) provide us with recursions for the three sequences $\{x_k, y_k, u_k\}$. In fact, one can eliminate the y and u variables to obtain the second-order recursion in x alone:

$$\frac{m(x_{k+1}) - m(x_k)}{x_k - x_{k-1}} = m'(x_k) \quad (k = 1, 2, ..., n), \qquad (2.35)$$

where

$$m(x_k) = \frac{1}{h(x_k)}. \tag{2.36}$$

For definiteness, consider the linear case $h(x) = \alpha x$. It is found that (2.35) then reduces to

$$x_k^2 = x_{k-1} x_{k+1}, \tag{2.37}$$

with solution

$$x_k = \beta \gamma^k. \tag{2.38}$$

The constants β, γ are to be determined from (2.30) and the fact that x_0 is prescribed. Setting (2.38) in (2.29), one sees that the optimal value of wash-flow in extractor k is

$$u_k = \frac{Q(1-\gamma)}{\alpha \gamma}, \tag{2.39}$$

and so independent of k. That is, in the case of linear h the optimal solution is to split the available wash-water equally between all n extractors, so that the solute concentration falls off geometrically along the extractor line, as indicated in (2.38).

If h is nonlinear, the second-order recursion (2.35) takes no very amenable form. The simplest course is then to regard k as a continuous variable, and approximate (2.35) by a differential equation. That is, it is virtually supposed that there are many extractors, each individually removing only a small proportion of the solute. Expanding $x_{k\pm1}$ in a Taylor series about the argument value k, one finds that (2.35) has the differential approximation

$$\frac{m''x'}{m'} + \frac{2x''}{x'} \sim 0, \tag{2.40}$$

where

$$x' = \frac{dx_k}{dk}, \quad m' = \frac{dm(x_k)}{dx_k},$$

etc. Equation (2.40) integrates immediately to yield the inverted form of solution

$$\int_{x_0}^{x_k} [-m'(\zeta)]^{\frac{1}{2}} d\zeta \sim \theta k, \tag{2.41}$$

where θ is a positive constant of integration, to be determined from condition (2.30).

So, for example, if

$$h(x) = \alpha x^{\nu}, \tag{2.42}$$

where ν is a positive index, then (2.41) yields

$$x_k \sim \left[x_0^{\frac{1}{2}(1-\nu)} - \left(\frac{\theta(1-\nu)}{2} \left(\frac{\alpha}{\nu} \right)^{\frac{1}{2}} \right) k \right]^{2/(1-\nu)}, \tag{2.43}$$

whence

$$u_k \sim -\frac{Qx'}{h(x)} \sim \frac{Q\theta}{\alpha\nu} \left[x_0^{\frac{1}{2}(1-\nu)} - \left(\frac{\theta(1-\nu)}{2} \left(\frac{\alpha}{\nu} \right)^{\frac{1}{2}} \right) k \right]. \tag{2.44}$$

Thus, u_k varies approximately linearly with k. If $\nu \geqslant 1$ then x_k and u_k never adopt zero values, but if $\nu < 1$ they may do so. In fact, the case $\nu < 1$ is physically unacceptable, for in this case $h(x)/x$ tends to infinity as x tends to zero: i.e. the proportion of solute removed in a single pass tends to unity as concentrations tend to zero. This would presumably never be the case. If it were, complete extraction could be achieved with a "continuum" of extractors and a sufficient (but finite) amount of wash-water. The previous proof of the strict positivity of u fails in such a case, because h is then not differentiable at the origin. The case $\nu > 1$ is that in which extraction efficiency increases with increasing concentration: this is straightforward.

The weak Lagrangian principle is valid, but not, in general, the strong one (see Exercise (8.6)). For example, the stationary point with respect to x_k located in (2.32) will not in general correspond to an absolute minimum of the form with respect to x_k. This certainly does not put the method out of court, because it does successfully determine the solution to a problem of practical interest.

2.6 MORE GENERAL CONSTRAINTS

It has already been remarked that types of constraint other than the equality (2.1) present themselves. For instance, the inequality (2.2) is the natural constraint for the allocation problem.

An inclusive reformulation of the problem is: maximize $f(x)$ in $\mathcal{X}$ subject to

$$b - g(x) \in \mathcal{C}, \tag{2.45}$$

where $\mathcal{C}$ is a set in R^m. If $\mathcal{C}$ consists only of the origin then the constraint reduces to the equality (2.1); if $\mathcal{C} = R_+^m$ then it reduces to the inequality (2.2). This is then a fairly flexible formulation of the problem, which will be denoted by $P_{\mathcal{C}}(b)$. The case of equality constraints will then be written $P_0(b)$.

That Lagrangian methods are still likely to be applicable follows from the fact that $P_{\mathcal{C}}(b)$ can be rewritten as an equality constraint problem: to maximize $f(x)$ with respect to x and z subject to

$$g(x) + z = b \tag{2.46}$$

and $(x, z) \in (\mathscr{X}, \mathscr{C})$. The new variable z has the appropriate name "slack variable". For example, for the allocation problem with constraint (2.2), z is just the vector of amounts of resources not used.

In a certain sense, then, the problem $P_\mathscr{C}(b)$ is no more general than the equality constraint problem, $P_0(b)$. However, the slack variable does enter the problem in a particular way: linearly in the constraints, and not at all in f. If one does not lose sight of the differing roles of the two arguments x and z, then the special structure of $P_\mathscr{C}(b)$ is preserved.

The presence of the slack variable gives an additional freedom of variation when optimizing, which makes the solution all the more special, as we shall see.

Because $P_\mathscr{C}(b)$ can be reduced to the equality constraint case, the geometric treatment of Section (2.3) carries over directly. The set of attainable values is now

$$\mathscr{B} = g(\mathscr{X}) + \mathscr{C}, \tag{2.47}$$

and the set of attainable (g, f) pairs is

$$\mathscr{D} = (g(\mathscr{X}) + \mathscr{C}, f(\mathscr{X})). \tag{2.48}$$

Theorems (2.1) and (2.2) can be restated as follows, no new proof being necessary. $\bar{z}$ is used to denote $b - g(\bar{x})$, where $\bar{x}$ is a solution to $P_\mathscr{C}(b)$.

Theorem (2.5)

(i) *Suppose that a supporting hyperplane to $\mathscr{D}$ exists at $(b, U(b))$. Then there exists a vector y and a non-negative scalar w such that $wf(x) - y^{\mathrm{T}}(gx) + z)$ is maximal for $(x, z) \in (\mathscr{X}, \mathscr{C})$ at $x = \bar{x}$, $z = \bar{z}$, the solution to $P_\mathscr{C}(b)$. Conversely, such a plane exists if a form with this maximal property exists.*

(ii), (iii) *As in Theorem (2.1).*

Theorem (2.6)

Suppose that a tangent plane to $\mathscr{D}$ exists at $(b, U(b))$, and that the derivatives f_x, g_x exist at $\bar{x}$. Then a vector y and a non-negative scalar w exist such that

$$(wf_x - y^{\mathrm{T}} g_x) s \leqslant y^{\mathrm{T}} t \tag{2.49}$$

for all directions (s, t) from $(\bar{x}, \bar{z})$ into the interior of $(\mathscr{X}, \mathscr{C})$. If U_b exists then one can normalize w to unity, and y^{T} then necessarily equals U_b.

The presence of the new variable z (or t) in the Lagrangian forms corresponds to the extra freedom of variation now available. It also implies conditions on the value of the multiplier y.

Theorem (2.7)

(i) *If $P_\mathscr{C}(b)$ is soluble by strong Lagrangian methods then*

$$y^{\mathrm{T}}(z - \bar{z}) \geqslant 0 \tag{2.50}$$

for any z in $\mathscr{C}$.

(ii) *If $P_{\mathscr{C}}(b)$ is soluble by weak Lagrangian methods, then*

$$y^\mathrm{T} t \geqslant 0 \qquad (2.51)$$

for any direction t from $\bar{z}$ into the interior of $\mathscr{C}$.

(iii) *If $\mathscr{C}$ is a cone, then, in either case,*

$$y^\mathrm{T} \bar{z} = 0, \qquad (2.52)$$

and, if $\mathscr{C}$ is a convex cone, then y also belongs to the conjugate cone $\mathscr{C}^*$.

Inequality (2.51) follows when $s = 0$ in (2.49). Inequality (2.50) correspondingly follows from the fact that $(\bar{x}, \bar{z})$ maximizes $wf(x) - y^\mathrm{T}(g(x) + z)$ in $(\mathscr{X}, \mathscr{C})$, so that $\bar{z}$ maximizes $-y^\mathrm{T} z$ in $\mathscr{C}$.

A *cone* is a set closed under dilations about the origin: if z belongs to it, then so does λz, for any non-negative scalar λ. The cone is convex if it is also closed under summation of its elements (see after the lemma in Section (2.4)). The *conjugate* of a cone $\mathscr{C}$ is the set of vectors y such that

$$y^\mathrm{T} z \geqslant 0 \qquad (2.53)$$

for all z in $\mathscr{C}$ (see Section (3.4)).

The last part of the theorem will be proved for the strong case only; the proof for the weak case is completely analogous. Set $z = \lambda \bar{z}$ in (2.50):

$$(\lambda - 1) y^\mathrm{T} \bar{z} \geqslant 0 \quad (\lambda > 0). \qquad (2.54)$$

This implies (2.52). If $\mathscr{C}$ is in addition convex, z can be chosen in (2.50) so that $z - \bar{z}$ is an arbitrary member of $\mathscr{C}$; this implies $y \in \mathscr{C}^*$.

The particular case of the allocation problem, with constraint $g(x) \leqslant b$, gives one a feeling for the assertions of Theorem (2.7). For this case the cone is the positive orthant R_+^m, a convex cone, and its conjugate is again R_+^m. The assertion $y \in \mathscr{C}^*$ then implies that

$$y \geqslant 0, \qquad (2.55)$$

i.e. that marginal prices for resources are positive. This is reasonable: if one is not compelled to use all available resources, then an increase in b implies merely an increase in the set $\mathscr{X}(b)$ of feasible solutions; and so an increase (not necessarily strict) in maximum return available. That is, $U(b)$ is non-decreasing in all arguments, and U_b, if it exists, is non-negative.

Condition (2.52) can be written

$$y^\mathrm{T}(b - g(\bar{x})) = 0, \qquad (2.56)$$

interpretable as a statement that resources unused in the optimal solution have zero value, when value is calculated on the basis of the effective marginal prices or "shadow" prices y. More particularly, since both vectors y and $b-g(\bar{x})$ are non-negative:

$$y_j = 0 \quad \text{if} \quad b_j - g_j(\bar{x}) > 0, \tag{2.57}$$

$$b_j - g_j(\bar{x}) \quad \text{if} \quad y_j > 0. \tag{2.58}$$

That is, a resource not fully utilized in the optimal solution has zero shadow price. Since additional amounts of such a resource could not increase the return, and would not be used, they would have no value. A resource with strictly positive shadow price is fully utilized.

Exercises

(2.20) Show that, if $U(b)$ is the maximum attained subject to (2.45), then

$$U(g(x)+z) \geqslant f(x)$$

for any (x, z) in $(\mathscr{X}, \mathscr{C})$, with equality for some such (x, z) for every attainable value of the argument of U.

(2.21) Hence demonstrate the inequality (2.49) with $w = 1$, $y^{\mathrm{T}} = U_b$, assuming existence of the appropriate derivatives.

(2.22) Consider the minimization of $f(x) = h(x_1) + h(x_2)$ in $x_1, x_2 \geqslant 0$ subject to $x_1 + x_2 \geqslant b$. Here

$$h(x) = \begin{cases} 0 & (x = 0), \\ x^2 - 2c_1 x + \frac{1}{2}c_2^2 & (x > 0), \end{cases}$$

where $c_2 > 2c_1 > 0$. Show that an optimal solution is

$$x_1 = x_2 = 0 \quad (b = 0),$$

$$x_1 = c_1, \quad x_2 = 0 \quad (0 < b \leqslant c_1),$$

$$x_1 = b, \quad x_2 = 0 \quad (c_1 < b \leqslant c_2),$$

$$x_1 = x_2 = b/2 \quad (b > c_2).$$

For example, $h(x)$ might represent the cost of producing a steam output x from a boiler, and b the total output required from a pair of similar boilers. The fact that $h(x)$ is *decreasing* in the range $(0 < x < c_1)$ represents the technical infeasibility of running a boiler at low positive rates.

(2.23) Find analogues of Theorems (2.3), (2.4) under the contraint (2.45).

Note. On p. 9 we defined a *feasible direction* for the unconstrained problem as a direction into the interior of $\mathscr{X}$. For the constrained problem it should perhaps be redefined as a direction into the interior of $\mathscr{X}(b)$, as some authors do, if consistency is to be maintained with the definition of a *feasible value* on p. 21. However, the former definition has been adhered to, and a feasible direction understood as one consistent with the constraint $x \in \mathscr{X}$ but not necessarily with others.

CHAPTER 3

The Strong Lagrangian Principle: Convexity

This chapter underpins the applications of following chapters. In it sufficient conditions are deduced for the set $\mathscr{D}$ of Section (2.3) to have the supporting hyperplane properties required there, and so for a strong Lagrangian principle to be valid. The case where such a hyperplane does not exist leads to the notion of a randomized solution. The concept of convexity proves central, and is developed from first principles in Sections 3.2, 3.4 and 3.5, although with an eye to the particular applications that will be made of it.

3.1 A SIMPLE SUFFICIENCY CONDITION; DUALITY; RANDOMIZED SOLUTIONS

For the strong principle there is an immediate and useful sufficiency condition which can be roughly expressed: "if the technique works, it is valid".

Theorem (3.1)

Suppose that a vector y can be found such that a value of x maximizing $f(x) - y^{\mathrm{T}} g(x)$ in $\mathscr{X}$, say $\bar{x}$, is feasible for $P_0(b)$. Then $\bar{x}$ solves $P_0(b)$.

Correspondingly, suppose that a y exists such that an (x, z) maximizing $f(x) - y^{\mathrm{T}}(g(x) + z)$ in $(\mathscr{X}, \mathscr{C})$, say $(\bar{x}, \bar{z})$, is feasible for $P_{\mathscr{C}}(b)$. Then $\bar{x}$ solves $P_{\mathscr{C}}(b)$.

Consider the first case: the second follows from it if $P_{\mathscr{C}}(b)$ is considered as a problem with the equality constraint (2.46). Since $\bar{x}$ is feasible for $P_0(b)$ then

$$f(\bar{x}) \leqslant U(b). \tag{3.1}$$

On the other hand,

$$U(b) = \max_{\mathcal{X}(b)} f = \max_{\mathcal{X}(b)} [f + y^{\mathrm{T}}(b - g)]$$

$$\leqslant \max_{\mathcal{X}} [f + y^{\mathrm{T}}(b - g)]$$

$$= f(\bar{x}) + y^{\mathrm{T}}(b - g(\bar{x}))$$

$$= f(\bar{x}). \tag{3.2}$$

It follows from (3.1) and (3.2) that equality holds in both relations and so the feasible value $\bar{x}$ solves $P_0(b)$.

As an example, consider the problem of Exercise (2.5). The Lagrangian form

$$\sum_k (x_k - x_k \log x_k - y_1 x_k - y_2 \varepsilon_k x_k) \tag{3.3}$$

has its unique maximum in $x \geqslant 0$ at the stationary point

$$\bar{x}_k = e^{-y_1 - y_2 \varepsilon_k}. \tag{3.4}$$

If $\varepsilon_1 < \varepsilon_2 < \varepsilon_3 < \dots$ and $\varepsilon_1 \leqslant \mathscr{E}/N < \lim \varepsilon_k$ then the multipliers y_1 and y_2 can certainly be chosen so that the two constraints of the exercise are satisfied; by the theorem thus proved $\bar{x}$ then solves the problem.

Another sufficient condition for solution of $P_\mathscr{C}(b)$ is the following:

Theorem (3.2)

Suppose that a vector y in $\mathscr{C}^*$ can be found such that a value of x maximizing $f(x) - y^{\mathrm{T}} g(x)$ in $\mathscr{X}$, say $\bar{x}$, is feasible for $P_\mathscr{C}(b)$ and satisfies

$$y^{\mathrm{T}}(b - g(\bar{x})) = 0. \tag{3.5}$$

Then $\bar{x}$ solves $P_\mathscr{C}(b)$. If $\mathscr{C}$ is a convex cone then this condition is equivalent to that in the second half of Theorem (3.1).

The proof that $\bar{x}$ is a solution runs very much as before: the sequences of relations (3.1) and (3.2) still hold as they stand. Proof of the last assertion is left to the reader.

The expression $f(x) - y^{\mathrm{T}} g(x)$ that has been considered is the x-dependent part of the form

$$L(x, y) = f(x) + y^{\mathrm{T}}(b - g(x)). \tag{3.6}$$

For many purposes this is the relevant quantity, and will be henceforth referred to as the *Lagrangian form*.

The steps up to the inequality in (3.2) can be written

$$U(b) \leqslant \max_{\mathcal{X}} L(x, y). \tag{3.7}$$

For $P_0(b)$ this inequality holds for any value of y; consequently,

$$U(b) \leqslant \min_{y} \max_{x \in \mathscr{X}} L(x, y). \tag{3.8}$$

The corresponding assertion for $P_{\mathscr{C}}(b)$ is

$$U(b) \leqslant \min_{y} \max_{\substack{x \in \mathscr{X} \\ z \in \mathscr{C}}} [f(x) + y^{\mathrm{T}}(b - g(x) - z)], \tag{3.9}$$

but either assertion can be written as the self-evident inequality

$$U(b) \leqslant \hat{U}(b) \overset{\text{def}}{=} \min_{y} \max_{\xi \in \mathscr{B}} [U(b) + y^{\mathrm{T}}(b - \xi)]. \tag{3.10}$$

The bound (3.9), or (3.10), is a significant one, in that its attainment is very nearly a necessary and sufficient condition for validity of the strong Lagrangian principle. That is, for a y to exist such that the (x, z) maximizing $f(x) + y^{\mathrm{T}}(b - g(x) - z)$ in $(\mathscr{X}, \mathscr{C})$ solves $P_{\mathscr{C}}(b)$, or, equivalently (see Theorem (2.1)), such that $U(\xi) - y^{\mathrm{T}} \xi$ is maximal in $\mathscr{B}$ at $\xi = b$.

Theorem (3.3)

  (i) $U(b)$ has the equivalent bounds asserted in (3.9) and (3.10).

  (ii) If $\mathscr{C}$ is a cone then the bound can be written

$$\hat{U}(b) = \min_{y \in \mathscr{C}^*} \max_{x \in \mathscr{X}} L(x, y). \tag{3.11}$$

  (iii) Equality holds in (3.10) if a y exists such that $U(\xi) - y^{\mathrm{T}} \xi$ is maximal in $\mathscr{B}$ at $\xi = b$.

  (iv) Such a y exists if equality holds in (3.10) and the minimizing y value is finite.

The validity of assertion (i) is evident. If $\mathscr{C}$ is a cone then the maximum in (3.9) will be $+\infty$ unless y belongs to $\mathscr{C}^*$, so the subsequent y minimization is effectively restricted to y in $\mathscr{C}^*$. Since then $y^{\mathrm{T}} z \geqslant 0$, the maximizing value of z will be such that $y^{\mathrm{T}} z = 0$, whence assertion (ii) follows.

It is assumed in (iii) that

$$U(b) - y^{\mathrm{T}} b = \max_{\xi \in \mathscr{B}} [U(\xi) - y^{\mathrm{T}} \xi], \tag{3.12}$$

which implies equality in (3.10). Conversely, the assumptions of (iv) imply the existence of a y such that (3.12) holds, with the conclusion that $U(\xi) - y^{\mathrm{T}} \xi$ is maximal in $\mathscr{B}$ at $\xi = b$.

The conclusions of the theorem are not very deep-lying. On the other hand, the interesting point does emerge that the appropriate Lagrangian multiplier y, if it exists, is the minimizing value in expressions (3.9) or (3.10). That is, the multiplier is the solution of a minimization problem (the so-called *dual problem*) associated with the original maximization problem. Moreover, the two problems share the common extremum $U(b)$ if the original (or primal) problem is amenable to strong Lagrangian methods. More specifically, the maximum attainable in the primal problem is $U(b)$, the minimum attainable in the dual problem is $\hat{U}(b)$; further, $U(b) \leqslant \hat{U}(b)$, with equality iff the strong Lagrangian principle is valid for the primal problem.

This question of duality will recur later. The natural immediate questions are: can the dual problem be given some intuitive meaning? and what interpretation can be attached to $\hat{U}(b)$ in the case $U(b) < \hat{U}(b)$, when Lagrangian methods fail?

The dual problem has an economic interpretation in certain cases (see Sections (3.7) and (4.3)); it also has an immediate geometric interpretation. Consider the graph of $U(\xi)$, i.e. the surface $\eta = U(\xi)$ in a space of $m+1$ dimensions with co-ordinates (ξ, η). Suppose that, for given y, the quantity $U(\xi) - y^{\mathrm{T}} \xi$ is maximal at ξ'. It can then be asserted that the hyperplane

$$\eta = U(\xi') + y^{\mathrm{T}}(\xi - \xi') \tag{3.13}$$

lies above the graph of $U(\xi)$; it is a supporting hyperplane to the set of points below this graph at the point $(\xi', U(\xi'))$. The quantity

$$U(\xi') + y^{\mathrm{T}}(b - \xi') = \max_{\xi} [U(\xi) + y^{\mathrm{T}}(b - \xi)] \tag{3.14}$$

is just the height of this plane above the coordinate hyperplane $\eta = 0$ at $\xi = b$; the length of the line PR in Figure (3.1). In the dual problem, the slope y of this supporting hyperplane is to be chosen so that PR is minimal. That is, the hyperplane is to be rolled over the graph of $U(\xi)$ until a position

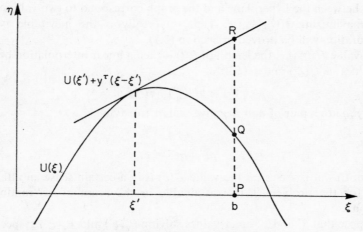

Figure 3.1 A geometric characterization of the dual problem

is reached at which PR is minimal. Evidently, since $PR \geqslant PQ$, PR will be minimized if the hyperplane can be rolled to Q so that the contact point has ξ-co-ordinate b, i.e. if a supporting hyperplane to the graph exists at $(b, U(b))$. If this is possible, then $PR = PQ$ and $\hat{U}(b) = U(b)$.

If this is not possible, then the configuration in which PR is minimal will be something of the nature illustrated in Figure (3.2). The case of scalar

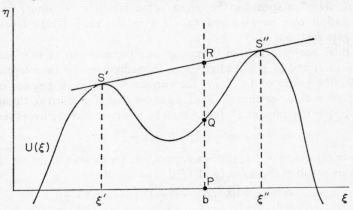

Figure 3.2 The geometric situation in the case where the graph of $U(\xi)$ does not possess a supporting hyperplane at the point $\xi = b$, and there is a "duality gap"

ξ has been illustrated ($m = 1$); the hyperplane for which PR is minimal bridges two local maxima of $U(\xi)$, between which b lies; the contact points S', S'' between the hyperplane and the graph correspond to two values ξ', ξ'', both maximizing $U(\xi) - y^{\mathrm{T}} \xi$. That this really is the inevitable type of configuration will be proved in Section (3.8).

The value $\hat{U}(b)$ (i.e. the length of PR) is then a linear interpolation between the values of $U(\xi')$ and $U(\xi'')$:

$$\hat{U}(b) = pU(\xi') + qU(\xi''), \tag{3.15}$$

where p, q are a pair of non-negative scalars satisfying

$$\begin{aligned} p + q &= 1, \\ p\xi' + q\xi'' &= b. \end{aligned} \tag{3.16}$$

From this, it is seen that the value $\hat{U}(b)$ *is* in a certain sense an attainable value for the problem $P_{\mathscr{C}}(b)$; attainable in the sense of a "randomized solution".

Suppose that x' and x'' are x-values solving $P_{\mathscr{C}}(\xi')$ and $P_{\mathscr{C}}(\xi'')$ respectively. Suppose that x is chosen equal to x' with probability p, and to x'' with probability q. Then, on account of (3.16), the constraint for $P_\mathscr{C}(b)$ (expressible as $\xi = b$, or $g(x) + z = b$), is fulfilled *on average*. The average value of f achieved is just $\hat{U}(b)$, by (3.15).

As an example, consider the case of a wholesaler who can achieve a gross rate of return $f(x)$ if he releases goods at rate x, where x is a scalar and the graph of $f(x)$ has the form of the curve in Figure (3.2). (A rather anomalous form, in this case, but the existence of two forms of goods-distribution, for example, might create a less favourable transition region for the release rate x.) Goods are supplied to him from the manufacturer at rate b: if he releases them at the same rate (corresponding to the constraint $x = b$) he will achieve a return $f(b)$. However, if he varies the release rate, taking it as x' a proportion p of the time and x'' a proportion q of the time, then he achieves the higher average rate of return $pf(x') + qf(x'')$. Yet, he releases goods on average at the same rate as they are supplied to him, so that there is no long-term accumulation or deficit.

These arguments will be formalized in Section (3.8), where it will be seen that the upper bound $\hat{U}(b)$ can always be attained (on average) if one is prepared to use randomized solutions, and that randomized solutions are necessary for this if $U(b) < \hat{U}(b)$.

Exercises

(3.1) Show that, if f is bounded above, then

$$\min_y L(x, y) = \begin{cases} f(x), & \text{if } g(x) = b, \\ -\infty, & \text{otherwise,} \end{cases}$$

so that, for $P_0(b)$,

$$U(b) = \max_x \min_y L(x, y).$$

The occurrence of equality in (3.8) thus amounts to commutativity of the two operations $\max\limits_x$ and $\min\limits_y$, when applied to $L(x, y)$. The celebrated *min–max theorem* gives general conditions for the validity of such a commutation, and will be studied in Chapter 10.

(3.2) Adapt Exercise (3.1) to the case of $P_\mathscr{C}(b)$, with $\mathscr{C}$ a cone.

(3.3) Show that $\hat{U}(b) = \min\limits_y \max\limits_{x \in \mathscr{X}} L(x, y)$ is also an upper bound to the maximum attainable for $P_0(b)$ by randomized solutions.

(3.4) Consider a boiler costing $h(x)$ per unit time to run at constant output rate x, where $h(x)$ has the form given in Exercise (2.22). Determine a method (randomized, if necessary) for running the boiler at a prescribed average output rate x for minimal average cost. What is the minimal cost as a function of x?

3.2 CONVEX SETS

The attempt to find a supporting hyperplane where none existed led, in the last section, to the notion of a randomized solution. A consideration of the same matters in reverse order leads, firstly, to the concept of convexity, secondly, to the celebrated "supporting hyperplane theorem".

Let ξ be the vector co-ordinate of a point in R^m, and $|\xi| = (\xi^T \xi)^{\frac{1}{2}}$ the distance of this point from the origin. Suppose that ξ belongs to a point set $\mathscr{A}$ in R^m. It is an *interior point* of $\mathscr{A}$ if $\mathscr{A}$ contains some spherical neighbourhood of ξ. It is a *boundary point* of $\mathscr{A}$ if all neighbourhoods of ξ contain points both of $\mathscr{A}$ and of its complement.

A set $\mathscr{A}$ of points in R^m is said to be *convex* if, for any two points ξ, ξ' belonging to $\mathscr{A}$, it is also true that $p\xi + q\xi'$ belongs to $\mathscr{A}$. (Recall the uniform convention, that p, q denote a pair of non-negative numbers adding to unity.) That is, $\mathscr{A}$ is such that the segment of straight line joining any two points of $\mathscr{A}$ lies wholly within $\mathscr{A}$. So, the set illustrated in (i) of Figure (3.3) is convex, that in (ii) is not.

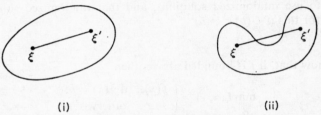

(i)   (ii)

Figure 3.3   Illustrations of convex and non-convex sets

It may be said that convexity corresponds to the requirement that $\mathscr{A}$ be *closed under averaging*; see Exercise (3.5). The notion applies equally well to sets defined in a linear topological space; a level of generality often needed (see Valentine (1964)), although not adopted here.

The *convex hull* of a set $\mathscr{A}$ is the smallest convex set which contains $\mathscr{A}$. In particular, if $\mathscr{A}$ consists of a finite number of points $\xi^1, \xi^2, ..., \xi^N$, then the convex hull of $\mathscr{A}$ is the set of points representable as an average of these:

$$\xi = \sum_{1}^{N} p_j \, \xi^j, \tag{3.17}$$

where $\{p_j\}$ constitutes a distribution (see Exercise (3.15)). This set constitutes a polyhedron in R^m; a convex polyhedron. If $\mathscr{A}$ is a set in the plane, its convex hull can be visualized as the set of points lying within a string pulled tight in a single loop around the points of $\mathscr{A}$.

A hyperplane in R^m,

$$\beta^T \xi = \gamma, \tag{3.18}$$

is a *supporting hyperplane* to $\mathscr{A}$ (where the set $\mathscr{A}$ is not necessarily convex, or even simply connected) if

$$\sup_{\xi \in \mathscr{A}} (\beta^T \xi) = \gamma. \tag{3.19}$$

That is, for points ξ of $\mathscr{A}$, the inequality

$$\beta^T \xi \leqslant \gamma \tag{3.20}$$

holds, with equality for at least one point, or limit point, of $\mathscr{A}$.

Normally $\mathscr{A}$ will be assumed closed, i.e. closed under the taking of limits. In such a case the supremum (3.19) is attained at a point of $\mathscr{A}$, say $\bar{\xi}$, and (3.19) can then be written as

$$\beta^T \xi \leqslant \beta^T \bar{\xi} \quad (\xi \in \mathscr{A}). \tag{3.21}$$

The point $\bar{\xi}$ thus maximizes the form $\beta^T \xi$ in $\mathscr{A}$. Because of this extremal characterization $\bar{\xi}$ is necessarily a boundary point of $\mathscr{A}$ ($\bar{\xi} + \delta\beta$ does not belong to $\mathscr{A}$ if $\delta > 0$), and the hyperplane (3.18), now written

$$\beta^T \xi = \beta^T \bar{\xi}, \tag{3.22}$$

would be referred to as a "supporting hyperplane to $\mathscr{A}$ at $\bar{\xi}$".

In Figure (3.4) some of the qualitative possibilities are illustrated: the stock example of (i), where the supporting hyperplane is also tangential; the case of polyhedral $\mathscr{A}$ in (ii), where the hyperplane may meet in a face rather than a single point; and the case of (iii), where the cusp admits infinitely many supporting hyperplanes. The following result is central.

Theorem (3.4)

Let $\mathscr{A}$ be convex, and ζ a point outside its closure. Then there is a hyperplane which separates ζ from $\mathscr{A}$.

The construction used is to take a hyperplane normal to the "perpendicular" from ζ to $\mathscr{A}$, intersecting this perpendicular at any point strictly between ζ and $\mathscr{A}$. $\mathscr{A}$ is supposed closed; if it is not, the following construction should be applied to its closure ($\mathscr{A}$ plus its limit points).

Let $\hat{\xi}$ be the element of $\mathscr{A}$ that minimizes $|\xi - \zeta|$, so that $\hat{\xi}$ is the "foot of the perpendicular". If ξ is an element of $\mathscr{A}$, and t a real number in $(0, 1)$, then $(1-t)\hat{\xi} + t\xi = \hat{\xi} + t(\xi - \hat{\xi})$ also belongs to $\mathscr{A}$, so that

$$|\hat{\xi} + t(\xi - \hat{\xi}) - \zeta|^2 \geqslant |\hat{\xi} - \zeta|^2,$$

or

$$2t(\hat{\xi} - \zeta)^T (\xi - \hat{\xi}) + t^2 |\xi - \hat{\xi}|^2 \geqslant 0.$$

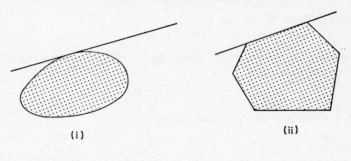

(i)   (ii)

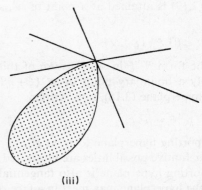

(iii)

Figure 3.4   Convex sets and supporting hyperplanes: some of the qualitative possibilities

Thus, in the limit of small t,

$$(\hat{\xi} - \zeta)^{\mathrm{T}} (\xi - \hat{\xi}) \geqslant 0 > -s |\hat{\xi} - \zeta|^2$$

for $s > 0$. Defining $\beta = \zeta - \hat{\xi}$, one has then

$$\beta^{\mathrm{T}} [\xi - s\zeta - (1-s)\hat{\xi}] < 0$$

for ξ in $\mathscr{A}$, so that the hyperplane

$$\beta^{\mathrm{T}} [\xi - s\zeta - (1-s)\hat{\xi}] = 0 \qquad\qquad (3.23)$$

does not intersect $\mathscr{A}$. The value of the linear form in the left-hand member of (3.23) at $\xi = \zeta$ is $(1-s)|\zeta - \hat{\xi}|^2$, which is positive if $s < 1$. The plane (3.23) thus separates ζ and $\mathscr{A}$ if $0 < s < 1$.

The classic "separating hyperplane theorem" is a slight generalization of Theorem (3.4); the adaptation is left to the reader in Exercise (3.12). A closely related and equally classic theorem follows.

Theorem (3.5) (*The supporting hyperplane theorem*)

A convex set has a supporting hyperplane at any point on its boundary.

Let the boundary point be ξ_0. The technique of proof is to consider a sequence of points ζ external to the set ($\mathscr{A}$, say) which converges to ξ_0. A hyperplane separating ζ and $\mathscr{A}$ exists at all stages; in the limit this must be a supporting hyperplane at ξ_0.

In order to formalize the argument the following lemma is needed:

Lemma

Let ξ_0 be a boundary point of a convex set $\mathscr{A}$. Then there exists a direction from ξ_0 strictly to the exterior of $\mathscr{A}$, in that there exists a vector σ such that $\xi_0 + \varepsilon\sigma$ does not belong to $\mathscr{A}$ for any $\varepsilon > 0$.

Suppose there exists no such direction. Then there exists an $\varepsilon > 0$ such that the 2^m points with co-ordinates $(\xi_{01} \pm \varepsilon, \xi_{02} \pm \varepsilon, \ldots, \xi_{0m} \pm \varepsilon)$ all belong to $\mathscr{A}$. But since $\mathscr{A}$ is convex, then the convex hull of these points belongs to $\mathscr{A}$: i.e. the hypercube centred on ξ_0 with these points as vertices. Thus a neighbourhood of ξ_0 belongs to $\mathscr{A}$, and ξ_0 is an interior point of $\mathscr{A}$. Since this contradicts hypothesis, the lemma is proven.

To proceed with the proof of the theorem, let σ be a direction to the exterior of $\mathscr{A}$ from ξ_0, and consider the sequence of points $\zeta_\varepsilon = \xi_0 + \varepsilon\sigma$ as $\varepsilon \downarrow 0$ through some sequence of values $\{\varepsilon_i\}$. For each ε a set of hyperplanes

$$\beta^T \xi = \gamma$$

exists, these hyperplanes separating ζ_ε and $\mathscr{A}$ in the non-strict sense that

$$\left. \begin{array}{l} \beta^T \zeta_\varepsilon \geqslant 0, \\ \beta^T \xi \leqslant 0 \quad (\xi \in \mathscr{A}). \end{array} \right\} \tag{3.24}$$

Let $\mathscr{H}_\varepsilon$ denote the set of such β, γ under the normalization $\beta^T \beta = 1$. It is then plain that $\mathscr{H}_\varepsilon$ is bounded (because of the normalization) and closed (because equality is permitted in (3.24)). Also $\mathscr{H}_\varepsilon \supset \mathscr{H}_{\varepsilon'}$ for $\varepsilon > \varepsilon'$, since a hyperplane separating $\mathscr{A}$ from $\zeta_{\varepsilon'}$ will also separate it from ζ_ε. The sequence $\{\mathscr{H}_{\varepsilon_i}\}$ is thus a monotone sequence of bounded, closed, non-empty sets, and will consequently have a non-empty limit, $\mathscr{H}_0$. (By the Heine–Borel theorem. See Apostol (p. 53, Theorem 3-37) for a version of the theorem immediately applicable to this case.) That is, at least one supporting hyperplane exists at ξ_0.

A special form of the theorem which is occasionally useful is

Theorem (3.6)

Let $\mathscr{A}$ be a convex polyhedron, ξ_0 a point on its boundary, and σ a direction such that $\xi_0 + \varepsilon\sigma$ is strictly exterior to $\mathscr{A}$ if $\varepsilon > 0$. Then a supporting hyperplane to $\mathscr{A}$ at ξ_0 exists which separates $\mathscr{A}$ strictly from $\xi_0 + \varepsilon\sigma$ for any $\varepsilon > 0$.

Let $\hat{\xi}_\varepsilon$ be the point in $\mathcal{A}$ nearest to $\xi_0 + \varepsilon\sigma$; use the construction of Theorem (3.4) to draw the hyperplane

$$\beta_\varepsilon^T(\xi - \hat{\xi}_\varepsilon) = 0 \tag{3.25}$$

through $\hat{\xi}_\varepsilon$ which separates $\mathcal{A}$ strictly from $\xi_0 + \varepsilon\sigma$, so that

$$\beta_\varepsilon^T(\xi - \hat{\xi}_\varepsilon) \leqslant 0 \quad (\xi \in \mathcal{A}), \tag{3.26}$$

$$\beta_\varepsilon^T(\xi_0 + \varepsilon\sigma - \hat{\xi}_\varepsilon) > 0. \tag{3.27}$$

Now, as $\varepsilon \downarrow 0$ then $\hat{\xi}_\varepsilon \to \xi_0$. For ε sufficiently small (but still strictly positive) ξ_0 will lie on some face of $\mathcal{A}$ on which $\hat{\xi}_\varepsilon$ also lies, and the point $\xi = \hat{\xi}_\varepsilon + \delta(\hat{\xi}_\varepsilon - \xi_0)$ will also belong to $\mathcal{A}$, for a range of δ which includes strictly positive and negative values. Requiring condition (3.26) for these values of ξ, one deduces that

$$\beta_\varepsilon^T(\xi_0 - \hat{\xi}_\varepsilon) = 0. \tag{3.28}$$

Thus the hyperplane (3.25) passes through ξ_0; since it separates $\mathcal{A}$ from an exterior point it is then a supporting hyperplane to $\mathcal{A}$ at ξ_0. The separation is strict, and condition (3.27) can be written

$$\beta_\varepsilon^T \sigma > 0. \tag{3.29}$$

The hyperplane (3.25) thus strictly separates $\xi_0 + \varepsilon'\sigma$ from $\mathcal{A}$ for any $\varepsilon' > 0$.

Exercises

(3.5) Show that if $\xi^1, \xi^2, \ldots, \xi^N$ belong to a given convex set, then so does any average $\sum p_j \xi^j$ of them.

(3.6) Consider a collection $\{\mathcal{A}_\nu\}$ of convex sets in a common space. Show that their intersection (i.e. the common part of all the $\mathcal{A}_\nu$) is convex.

(3.7) Hence a set which is the intersection of a number of half-spaces

$$\beta_\nu^T \xi \leqslant \gamma_\nu \quad \text{(for all } \nu \text{ in some set)}$$

is convex. By definition, a *convex polyhedron* if the number of ν-values is finite.

(3.8) A convex polyhedron can be defined as the intersection of a finite number of half-spaces, or as the set of points representable as in Exercise (3.5) for fixed $\xi^1, \xi^2, \ldots, \xi^N$ and a variable distribution $\{p_j\}$. Prove the equivalence of these two definitions.

(3.9) Consider a collection of sets $\{\mathcal{A}_\nu\}$ in a common space. Suppose that ν takes values in a convex set, and that $p\xi + q\xi'$ belongs to $\mathcal{A}_{p\nu + q\nu'}$ if $\xi \in \mathcal{A}_\nu$ and $\xi' \in \mathcal{A}_{\nu'}$. Show that the union of the $\mathcal{A}_\nu$ is convex.

(3.10) Show that, if $\mathscr{A}$ is convex, then so is its closure.

(3.11) Show by an example that the convex hull of $\mathscr{A}$ may be larger than the set of points $p\xi + q\xi'$ for ξ, ξ' in $\mathscr{A}$ and (p, q) a distribution.

(3.12) *The separating hyperplane theorem.* This name is usually given to a slightly more general result than Theorem (3.4); to the statement that there exists a hyperplane separating two disjoint convex sets. More specifically, suppose that sets $\mathscr{A}, \mathscr{A}'$ are each convex, and are separated by a positive amount, so that $\delta = \inf |\xi - \xi'| > 0$ the infimum being taken over ξ in $\mathscr{A}$ and ξ' in $\mathscr{A}'$. Suppose also that this infimum is achieved for finite ξ, ξ'. Adapt Theorem (3.4) to show that a hyperplane $\beta^T \xi = \gamma$ exists which strictly separates the two sets, i.e. such that

$$\beta^T \xi < \gamma \quad (\xi \in \mathscr{A}),$$

$$\beta^T \xi > \gamma \quad (\xi \in \mathscr{A}').$$

(3.13) Consider the convex hull $[\mathscr{A}]$ of a closed point set $\mathscr{A}$ in R^m. Show that any point of $\mathscr{A}$ possessing a supporting hyperplane is a boundary point of $[\mathscr{A}]$. Furthermore, show that any supporting hyperplane to $[\mathscr{A}]$ contains points of $\mathscr{A}$.

(3.14) Consider a particular supporting hyperplane to $[\mathscr{A}]$, and denote the set of all points in this hyperplane by $\mathscr{S}$. Show, by proving an inclusion relationship each way, that

$$\mathscr{S} \cap [\mathscr{A}] = [\mathscr{S} \cap \mathscr{A}]$$

(3.15) Continuing from Exercises (3.13) and (3.14), show that any point of $[\mathscr{A}]$ can be represented as an average of at most $m+1$ points of $\mathscr{A}$, at least if $\mathscr{A}$ is bounded. (Let ξ be the point, and ξ' any point of $\mathscr{A}$. Join ξ' to ξ by a straight line, and continue the line through ξ until it meets the boundary of $[\mathscr{A}]$ in a point ξ'', so that ξ is an average of ξ' and ξ''. By Exercise (3.14), ξ'' lies in the convex hull of those points of $\mathscr{A}$ which lie in the supporting hyperplane to $[\mathscr{A}]$ at ξ'': a set in R^{m-1}. This fact provides the basis for an induction on m.)

(3.16) It is natural to ask whether Theorem (3.5) has a converse, i.e. whether convexity is a necessary as well as a sufficient condition for a set to have a supporting hyperplane at all its boundary points. This is not quite true: consider a set of just two points (or, more generally, a set of points which are all boundary points of some convex set). Show that a set possesses a supporting hyperplane at all its boundary points if and only if these are also boundary points of some convex set.

3.3 APPLICATIONS OF THE SUPPORTING HYPERPLANE THEOREM

The most important application of the supporting hyperplane theorem, Theorem (3.5), will certainly be the conclusion that, if the set $\mathscr{D} = (\mathscr{B}, f(\mathscr{X}))$ of Sections (2.2) and (2.6) is convex, then it will possess supporting hyperplanes, and so, by Theorem (2.1), a strong Lagrangian principle will be valid. This is a genuinely useful condition for validity of the principle, because conditions on f, g, $\mathscr{X}$ and $\mathscr{C}$ which ensure convexity of $\mathscr{D}$ are easily established: this point is pursued in Section (3.6). Furthermore, the condition is not far from being a necessary one if the principle is to hold for all b in $\mathscr{B}$, in that $U(b)$ must then coincide with its least convex majorant in $\mathscr{B}$; see Section (3.8).

Actually, seeing that only the upper boundary of $\mathscr{D}$ is of interest, and the existence of supporting hyperplanes is required only to points on this upper boundary, it is sufficient if the set $\mathscr{D}_+ = (\mathscr{B}, f(\mathscr{X}) - R_+')$ is convex. That is, the set of points (ξ, η) satisfying $\xi \in \mathscr{B}$, $\eta \leqslant U(\xi)$. The relation between $\mathscr{D}$ and $\mathscr{D}_+$ is sketched in Figure (3.5).

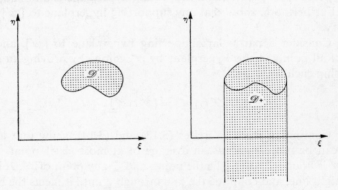

Figure 3.5 The relation between the sets $\mathscr{D}$ and $\mathscr{D}_+$

Another application of the supporting hyperplane theorem is to prove the lemma of Section (2.4), which might be rephrased in a neutral notation as follows: Suppose that $c^T x \leqslant 0$ if $Ax \leqslant 0$ and $x \in \mathscr{K}$, where x and c are n-vectors, A an $m \times n$ matrix, and $\mathscr{K}$ a convex cone in R^n. Then there exists an m-vector y and a non-negative scalar w such that

$$wc^T x - y^T Ax \leqslant 0 \tag{3.30}$$

for x in $\mathscr{K}$. The constant w can be normalized to unity if $\mathscr{K}$ is polyhedral. (The validity of condition (i) of the lemma is self-evident.)

Consider the (ξ, η) set $\mathscr{A}$ in R^{m+1} with elements

$$\left.\begin{aligned} \xi &= Ax, \\ \eta &= c^{\mathrm{T}} x \end{aligned}\right\} \tag{3.31}$$

for some x in $\mathscr{K}$. Then $\mathscr{A}$ is plainly convex, and it has the origin as a boundary point. There will consequently be a supporting hyperplane at the origin; i.e. a w and y such that

$$w\eta - y^{\mathrm{T}} \xi \leqslant 0 \tag{3.32}$$

for $(\xi, \eta) \in \mathscr{A}$. The relation (3.30) follows from (3.31) and (3.32). The point $(0, \delta)$ does not belong to $\mathscr{A}$ if $\delta > 0$; by taking the supporting hyperplane as a limit $(\delta \downarrow 0)$ of a hyperplan eseparating $\mathscr{A}$ from $(0, \delta)$, one sees that $w\delta \geqslant 0$, or $w \geqslant 0$.

As usual, the exclusion of the possibility $w = 0$ is the hardest part. Suppose that $\mathscr{K}$ is polyhedral. Define the set $\mathscr{A}_M$ whose elements (ξ, η) satisfy (3.31) for some x in the intersection of $\mathscr{K}$ and the hypercube

$$-M \leqslant x_k \leqslant M \quad (k = 1, 2, \dots, n).$$

x is thus restricted to lying in a bounded convex polyhedron, and $\mathscr{A}_M$ will also be a bounded convex polyhedron in R^{m+1}, with the origin as boundary point. By Theorem (3.6) there is a supporting hyperplane $w\eta - y^{\mathrm{T}} \xi = 0$ at the origin which separates $\mathscr{A}$ strictly from $(0, \delta)$ for $\delta > 0$, so that (3.32) holds, and also $w\delta > 0$, or $w > 0$.

The elements of $\mathscr{A}$ are derived from those of $\mathscr{A}_M$ by the expansion of scale $M \to +\infty$, and can all be represented in the form $(\lambda\xi, \lambda\eta)$ for λ a non-negative scalar, and (ξ, η) an element of $\mathscr{A}_M$. A relation (3.32) which holds for all elements of $\mathscr{A}_M$ thus holds also for all elements of $\mathscr{A}$. That is, there is a relation (3.30) with $w > 0$ which holds for all x in $\mathscr{K}$.

3.4 CONES AND THEIR CONJUGATES

Some definitions already given in Sections (2.4) and (2.6) are recalled. A cone (with vertex at the origin, understood) is a set $\mathscr{C}$ such that $\lambda\xi$ belongs to if ξ does, for any non-negative scalar λ. That is, a cone is a set *closed under dilations about the vertex point*. A *convex cone* is then a set closed under both dilation and averaging: if ξ, ξ' belong to it, then so does $\lambda\xi + \mu\xi'$ for any non-negative λ, μ.

The *conjugate* or *dual* of a cone $\mathscr{C}$, denoted $\mathscr{C}^*$, is the set of vectors β such that

$$\beta^{\mathrm{T}} \xi \geqslant 0 \tag{3.33}$$

for all ξ in $\mathscr{C}$. For example, suppose that $\mathscr{C}$ is just the origin, the single point $\xi = 0$. Then (3.33) is satisfied for any β, so that $\mathscr{C}^*$ is the whole space of β. If, on the other hand, $\mathscr{C}$ is the positive orthant, $\xi \geqslant 0$, then $\mathscr{C}^*$ is also the positive orthant, $\beta \geqslant 0$, in β-space. These two examples illustrate a general point: the larger $\mathscr{C}$, the smaller $\mathscr{C}^*$, because (3.33) is then more demanding on β.

Exercises

(3.17) Suppose that $\mathscr{A}, \mathscr{C}$ are cones. Show that
  (i) $\mathscr{A} \subset \mathscr{C}$ implies that $\mathscr{A}^* \supset \mathscr{C}^*$;
  (ii) $\mathscr{A}^{**}$ is the convex hull of $\mathscr{A}$;
  (iii) $(\mathscr{A} + \mathscr{C})^* = \mathscr{A}^* \cap \mathscr{C}^*$ if $\mathscr{A}, \mathscr{C}$ both contain the origin.

(3.18) Suppose that $\mathscr{A}, \mathscr{C}$ are closed convex cones. By appealing to Exercise (3.17) and the supporting hyperplane theorem, show that
  (i) $\mathscr{A} = \mathscr{A}^{**}$;
  (ii) $(\mathscr{A} \cap \mathscr{C})^*$ is the closure of $\mathscr{A}^* + \mathscr{C}^*$.

(3.19) Note that the Farkas lemma proved in Section (3.3) is a special case of Exercise (3.18) (ii).

3.5 CONVEX FUNCTIONS

Consider a real scalar-valued function $f(x)$ defined on a set $\mathscr{X}$ in R^n. The function is said to be convex if it has the property

$$f(px + qx') \leqslant pf(x) + qf(x') \tag{3.34}$$

for any x, x' in $\mathscr{X}$, and any pair of non-negative scalars p, q adding to unity. Implicit in the statement is the condition that $px + qx'$ also belongs to $\mathscr{X}$, i.e. that the set $\mathscr{X}$ is convex.

Condition (3.34) states that, if the graph of $f(x)$, say $t = f(x)$, is drawn in an $(n+1)$-dimensional space, then the line segment joining two points on the graph lies wholly above the graph. In other words, the set of points lying above the graph of $f(x)$ for x in $\mathscr{X}$ is convex: the shaded set in Figure (3.6). This observation gives a basis for a more formal definition, which relates convexity of functions to convexity of sets. The *function $f(x)$ is said to be convex* in $\mathscr{X}$ if the *set* $(\mathscr{X}, f(\mathscr{X}) + R_+^1)$ in R^{n+1} is convex.

The function $f(x)$ is *concave* if $-f(x)$ is convex. A concave function thus obeys the reverse inequality to (3.34):

$$f(px + qx') \geqslant pf(x) + qf(x'), \tag{3.35}$$

and it is the set of points *below* its graph, $(\mathscr{X}, f(\mathscr{X}) - R_+^1)$, which is convex.

Obviously all discussion of concave functions can immediately be phrased in terms of convex functions. On the other hand, concavity is a very natural property in maximization problems (as is convexity in minimization problems), so the argument will usually be put for concave rather than convex functions.

For example, suppose that $f(x)$ represents the utility derived from a vector investment x (the different components of x representing different types of investment). A utility is naturally a quantity that one wishes to maximize.

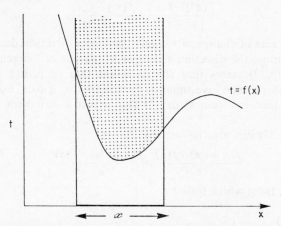

Figure 3.6   The function $f(x)$ is convex in $\mathcal{X}$ if the set above the graph of $f(x)$ for x in $\mathcal{X}$ (shaded) is convex

Then the concavity condition (3.35) expresses three properties which are very characteristic of utility functions: the desirability of diversity, the desirability of certainty and a law of decreasing returns. Relation (3.35) states that the mixed investment $px+qx'$ has a utility not less in value than the averaged utility of the two extreme investments x, x': the utility of possessing an orange and an apple is greater (or, at least, not less) than the mean utility of possessing just two oranges or just two apples. Similarly, the return from a certain investment $px+qx'$ is not less than the expected return from an uncertain situation in which the investment may be x with probability p, or x' with probability q.

To see a law of decreasing returns, let x, x', x'' be three collinear points in order, so that if s is the direction of the line then

$$\left. \begin{array}{l} x' = x + \varepsilon's, \\ x'' = x + \varepsilon''s \end{array} \right\} \tag{3.36}$$

with $0 < \varepsilon' < \varepsilon''$. Inequality (3.35) then implies that

$$f(x') \geqslant pf(x) + qf(x'') \tag{3.37}$$

3

with

$$p = 1 - \frac{\varepsilon'}{\varepsilon''} = \frac{|x'' - x'|}{|x'' - x|}. \qquad (3.38)$$

(3.37) can be rewritten as

$$\frac{f(x'') - f(x')}{|x'' - x'|} \leqslant \frac{f(x') - f(x)}{|x' - x|}. \qquad (3.39)$$

In words: the rate of change of f in any particular direction decreases in that direction. Suppose the direction s is regarded as one of "increase" in investment. Then (3.39) states that the marginal rate of return (proportional increase in utility) when investment is increased from x to x' exceeds that for a subsequent increase from x' to x''. This is just a statement of decreasing returns.

Relation (3.37) can also be written

$$\frac{f(x + \varepsilon' s) - f(x)}{\varepsilon'} \geqslant \frac{f(x + \varepsilon'' s) - f(x)}{\varepsilon''} \qquad (3.40)$$

for $0 < \varepsilon' < \varepsilon''$, from which follows

Theorem (3.7)

Suppose f concave in $\mathscr{X}$, x a point of $\mathscr{X}$ at which f is finite, and s a direction from x into the interior of $\mathscr{X}$. Then f possesses a directional derivative at x in direction s.

The derivative in direction s would be the limit of $[f(x + \varepsilon s) - f(x)]/\varepsilon$ as $\varepsilon \downarrow 0$. But (3.40) states that this quotient is monotone in ε, so the limit (possibly infinite) exists. The limit will be denoted by $D_s f(x)$. If the derivative f_x exists at x, then $D_s f$ necessarily equals $f_x s$. However, directional derivatives may well exist in the absence of a derivative. As seen in Figure (3.7) (see p. 61), the graph of $f(x)$ may have a discontinuity in slope at a particular point, although sloping away from this point in a perfectly smooth fashion in any given direction.

However, the principal property of a concave function in the optimization context must follow from the fact that, if $\bar{x}$ is a point of $\mathscr{X}$, then $(\bar{x}, f(\bar{x}))$ is a boundary point of the convex set $(\mathscr{X}, f(\mathscr{X}) - R_+^1)$, and, in virtue of Theorem (3.5), there will be a supporting hyperplane to the set at this point.

Theorem (3.8)

Suppose $\mathscr{X}$ convex, $f(x)$ concave in $\mathscr{X}$, and finite at the finite point $\bar{x}$ of $\mathscr{X}$. Then

(i) *A vector y and a non-negative scalar w exist such that $wf(x) - y^T x$ is maximal in $\mathscr{X}$ at $x = \bar{x}$.*

(ii) *The directional derivatives of f at $\bar{x}$ satisfy*

$$y^T s \geqslant w D_s f \tag{3.41}$$

for directions s from $\bar{x}$ into $\mathcal{X}$; and any (y, w) satisfying (3.41) and $w \geqslant 0$ have the property enunciated in (i).

(iii) *The constant w may be normalized to unity if either* (a) $\bar{x}$ *is interior to $\mathcal{X}$ and $f(x)$ finite in a neighbourhood $N(\bar{x})$ of $\bar{x}$, or* (b) $f(x)$ *is piecewise linear and finite in $N(\bar{x}) \cap \mathcal{X}$ or* (c) $D_s f$ *is finite at $\bar{x}$ for finite s directed from $\bar{x}$ into $\mathcal{X}$.*

(iv) *Suppose that $\bar{x}$ is interior to $\mathcal{X}$ and f_x exists there and is finite. If w is normalized to unity (as is possible, by (iii)) then y necessarily equals $f_x{}^T$.*

The existence of a supporting hyperplane to $(\mathcal{X}, f(\mathcal{X}) - R_+{}^1)$ at $(\bar{x}, f(\bar{x}))$ implies the existence of y, w such that

$$w f(\bar{x}) - y^T \bar{x} \geqslant w(f(x) - t) - y^T x \tag{3.42}$$

for $x \in \mathcal{X}, t \geqslant 0$. Setting $x = \bar{x}$, one deduces that $w \geqslant 0$; setting $t = 0$, one deduces that $w f(x) - y^T x$ is maximal in $\mathcal{X}$ at $\bar{x}$.

Setting $t = 0$, $x = \bar{x} + \varepsilon s$, in (3.42), and letting ε tend to zero from above, one deduces (3.41). To establish the converse result expressed in (ii), suppose that $x = \bar{x} + s$ belongs to $\mathcal{X}$. Letting $\varepsilon'' = 1$ and $\varepsilon' \downarrow 0$ in (3.40), one deduces that

$$D_s f(\bar{x}) \geqslant f(\bar{x} + s) - f(\bar{x}). \tag{3.43}$$

From (3.41) and (3.43) it follows that

$$y^T s \geqslant w[f(\bar{x} + s) - f(\bar{x})]$$

or that

$$w f(\bar{x}) - y^T \bar{x} \geqslant w f(x) - y^T x, \tag{3.44}$$

as is required. Assertions (iiia) and (iv) have already been established in Theorem (2.1). Assertion (iiib) follows from the fact that $(\bar{x}, f(\bar{x}) + \delta)$ is, for $\delta > 0$, strictly exterior to the locally polyhedral set $(\mathcal{X}, f(\mathcal{X}) - R_+{}^1)$. The supporting hyperplane at $(\bar{x}, f(\bar{x}))$ may be chosen to separate $(\bar{x}, f(\bar{x}) + \delta)$ strictly from this set, by Theorem (3.6), whence it follows that $w > 0$.

Assertion (iiic) includes (iiib), and provides the most substantial new feature of the theorem. It states that, if the directional derivatives of f at $\bar{x}$ are finite, then a supporting hyperplane to the graph of f at $\bar{x}$ which is not "vertical" can be found, even if $\bar{x}$ is on the boundary of $\mathcal{X}$. This is a useful conclusion, but one requiring some new arguments.

Consider the set of s directed into $\mathcal{X}$ from $\bar{x}$ (i.e. such that $\bar{x} + \varepsilon s$ belongs to $\mathcal{X}$ for all sufficiently small positive ε); denote this by $\mathcal{K}$. One readily verifies that $\mathcal{K}$ is convex. Regard $D_s f(\bar{x})$ as a function of s; say $\phi(s)$. By letting ε tend to zero in

$$f(\bar{x} + p\varepsilon s + q\varepsilon s') - f(\bar{x}) \geqslant p[f(\bar{x} + \varepsilon s) - f(x)] + q[f(\bar{x} + \varepsilon s') - f(\bar{x})] \qquad (3.45)$$

one deduces that $\phi(s)$ is concave. If $\phi(s)$ is finite for finite s in $\mathcal{K}$, and $\bar{s}$ interior to $\mathcal{K}$, it thus follows from (i) and (iiia) of the theorem that a vector y exists such that $\phi(s) - y^{\mathrm{T}} s$ is maximal in $\mathcal{K}$ at $s = \bar{s}$. Note now that

$$\phi(\lambda s) = \lambda \phi(s) \qquad (3.46)$$

for scalar λ, and hence that

$$\phi(\bar{s}) - y^{\mathrm{T}} \bar{s} \geqslant \phi(\lambda \bar{s}) - \lambda y^{\mathrm{T}} \bar{s} = \lambda[\phi(\bar{s}) - y^{\mathrm{T}} \bar{s}]. \qquad (3.47)$$

Setting $\lambda = 1 \pm \delta$ one deduces from (3.47) that $\phi(\bar{s}) - y^{\mathrm{T}} \bar{s} = 0$. That is, a vector y exists such that

$$\phi(s) - y^{\mathrm{T}} s \leqslant 0 \qquad (3.48)$$

or

$$y^{\mathrm{T}} s \geqslant D_s f(\bar{x}) \qquad (3.49)$$

for all s in $\mathcal{K}$. Thus (3.41) holds for this y with $w = 1$, and, by the second half of (ii), $f(x) - y^{\mathrm{T}} x$ is then maximal in $\mathcal{X}$ at $\bar{x}$. The theorem is thus proven.

Exercises

(3.20) *Jensen's inequality.* If f is convex then it follows from the characterizing property (3.34) that

$$f(Ex) \leqslant Ef(x)$$

where E is the expectation operator corresponding to a distribution over a finite number of points of $\mathcal{X}$.

Let E be the expectation operator for a general distribution over $\mathcal{X}$, characterized only by the properties of being linear, positive (i.e. $E[\psi(x)] \geqslant 0$ if $\psi(x) \geqslant 0$ in $\mathcal{X}$) and normalized so that $E(1) = 1$. Suppose $E(x)$ finite and f finite in a neighbourhood of $E(x)$. Then show, by appeal to the supporting hyperplane theorem, that the inequality above is still valid.

(3.21) Show that $f(x) = x \log x$ is convex in R_+^1. Hence show, by appeal to Jensen's inequality, that

$$\sum p_j \log(p_j/p_j') \geqslant 0$$

for $\{p_j\}, \{p_j'\}$ distributions on the same denumerable point set, and that equality holds iff $p_j = p_j'$ for all j in this set. These statements have implications in information theory; see the latter part of Section (10.7).

(3.22) Show that, if the functions $f_\nu(x)$ are convex, then so is

$$\bar{f}(x) = \max_\nu f_\nu(x).$$

Note that this follows from the statement: the intersection of a number of convex sets is convex. Are there conditions under which one could assert convexity of

$$\underline{f}(x) = \min_{\nu} f_{\nu}(x)?$$

(cf. Exercise (3.9)). In general, $\underline{f}$ is not convex, as almost any example will show.

(3.23) Suppose f convex in the convex set $\mathscr{X}$, and that it attains its supremum in $\mathscr{X}$. Show that this supremum is attained on the boundary of $\mathscr{X}$. Show that if f is strictly convex (i.e. strict inequality holds in (3.34), for $x \neq x', p \neq 0, 1$) then the supremum cannot be attained in the interior of $\mathscr{X}$. Note that the differing characters of the solution of Exercise (2.12) in the cases $\beta < 0, \beta > 0$ originated from the fact that f was respectively concave and convex in these two cases.

(3.24) Consider a differentiable function of a scalar variable, $f(x)$. Show that f is convex iff df/dx is a non-decreasing function of x.

(3.25) What function is both convex and concave?

(3.26) Suppose that f is convex, and possesses second-order derivatives at a point. Show that the matrix of these derivatives is non-negative definite.

(3.27) Show that, of the w, y satisfying (3.41) for all feasible s, there exists a w, y such that equality holds in (3.41) for any particular feasible s.

Suppose that w can be set equal to unity, and denote the set of y satisfying (3.41) by $\mathscr{Y}$. Show then that

$$D_s f = \min_{y \in \mathscr{Y}} (y^{\mathrm{T}} s)$$

(cf. the arguments used to establish (iiic) of Theorem (3.8)).

3.6 THE STRONG LAGRANGIAN PRINCIPLE

Some simple, verifiable conditions which ensure validity of the strong Lagrangian principle can now be stated. We begin with a portmanteau theorem, which collects a number of scattered assertions.

Recall the definition of the problem $P_{\mathscr{C}}(b)$: the maximization of $f(x)$ in $\mathscr{X}$ subject to

$$b - g(x) \in \mathscr{C}. \tag{3.50}$$

$U(b)$ denotes the maximum attained, and $\mathscr{B} = g(\mathscr{X}) + \mathscr{C}$ the set of attainable b. A solution to the problem is denoted by $x(b)$ or $\bar{x}$, as convenient. The vector $b - g(\bar{x})$ is denoted $\bar{z}$.

Theorem (3.9)

Suppose $\mathscr{B}$ convex, $U(\xi)$ concave in $\mathscr{B}$, and finite at the finite value $\xi = b$. Then

(i) *A vector y and non-negative scalar w exist such that $wU(\xi) - y^{\mathrm{T}}\xi$ is maximal in $\mathscr{B}$ at $\xi = b$.*

(ii) *The directional derivatives of U at b exist and satisfy*

$$y^{\mathrm{T}} s \geqslant w D_s U \tag{3.51}$$

for directions s from b into the interior of $\mathscr{B}$.

(iii) *The constant w may be normalized to unity if either* (a) *b is interior to $\mathscr{B}$ and $U(\xi)$ finite in a neighbourhood $N(b)$ of b or* (b) *$U(\xi)$ is piecewise linear and finite in $N(b) \cap \mathscr{B}$, or* (c) *$D_s U$ is finite at b for finite s directed from b into $\mathscr{B}$.*

(iv) *If b is interior to $\mathscr{B}$ and the derivative U_b exists and is finite, then w may be normalized to unity and y necessarily then equals $U_b{}^{\mathrm{T}}$.*

(v) *Assertion* (i) *implies that $wf(x) - y^{\mathrm{T}}(g(x) + z)$ is maximal for (x, z) in $(\mathscr{X}, \mathscr{C})$ at $(\bar{x}, \bar{z})$. In particular, $wf(x) - y^{\mathrm{T}}g(x)$ is maximal in $\mathscr{X}$ at $x = \bar{x}$.*

(vi) *If $\mathscr{C}$ is a convex cone, then* (v) *implies that $y \in \mathscr{C}^*$ and $y^{\mathrm{T}}(b - g(\bar{x})) = 0$.*

Assertions (i)–(iv) are lifted bodily from Theorem (3.8). Assertion (v) follows from (i), and emphasizes the strong Lagrangian principle; (vi) is proved in Theorem (2.7). The theorem thus adds nothing to earlier results, but merely summarizes the Lagrangian-type conclusions that follow from concavity of $U(\xi)$, plus various regularity conditions. Conditions ensuring this concavity are formulated in Theorem (3.10).

Inequality (3.51) extends the price characterization of the multiplier $y = U_b{}^{\mathrm{T}}$ to the case where U_b does not necessarily exist. To take the simplest example, suppose that b is scalar, so that the directional derivatives reduce to the right- and left-derivatives

$$\left.\begin{array}{l} \left(\dfrac{\partial U}{\partial b}\right)_+ = \lim_{\varepsilon \downarrow 0} \dfrac{U(b + \varepsilon) - U(b)}{\varepsilon}, \\[3mm] \left(\dfrac{\partial U}{\partial b}\right)_- = \lim_{\varepsilon \downarrow 0} \dfrac{U(b) - U(b - \varepsilon)}{\varepsilon}. \end{array}\right\} \tag{3.52}$$

These would then be "right" and "left" prices: the rate of change of return with amount of resource available (in the allocation case) according as the amount of resource is increased or decreased. The two coincide if and only if $\partial U / \partial b$ exists at b. If both directional derivatives are finite, then w can certainly be set equal to unity, and inequality (3.51) reduces to

$$\left(\frac{\partial U}{\partial b}\right)_- \geqslant y \geqslant \left(\frac{\partial U}{\partial b}\right)_+. \tag{3.53}$$

That is, the right-price does not exceed the left-price (decreasing returns again) and the slope of the supporting hyperplane at b may adopt any intermediate value, as is clear from Figure (3.7).

Some simple conditions which will ensure concavity of U, the principal assumption of Theorem (3.9), have yet to be found. To this end, consider a new concept: that the vector function $g(x)$ is $\mathscr{C}$-convex if

$$pg(x)+qg(x')-g(px+qx')\in\mathscr{C} \tag{3.54}$$

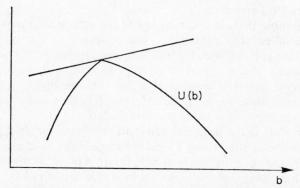

Figure 3.7 A concave function with discontinuous derivative. The derivative to left and right of the cusp defines a "left" and "right" price, respectively: the value of the slope of the supporting line at the cusp must lie between these two values

for x, x' in $\mathscr{X}$. This reduces to the notion of ordinary convexity in the case $\mathscr{C} = R_+^1$. In the case where $\mathscr{C}$ is the origin in R^m, (3.54) requires that $g(x)$ be linear in x.

Theorem (3.10)

Suppose $\mathscr{X}$ convex, $\mathscr{C}$ a convex cone, $f(x)$ concave, and $g(x)$ $\mathscr{C}$-convex. Then $\mathscr{B}$ is convex, and $U(b)$ concave in $\mathscr{B}$.

Consider a pair of b values, b' and b'', and a pair of x values, x' and x'', and set

$$b = pb'+qb'',$$

$$x = px'+qx''.$$

Suppose that x' and x'' are feasible for $P_{\mathscr{C}}(b')$ and $P_{\mathscr{C}}(b'')$ respectively. Then x belongs to $\mathscr{X}$, and

$$b-g(x) = p[b'-g(x')]+q[b''-g(x'')]+[pg(x')+qg(x'')-g(x)]$$

which belongs to $\mathscr{C}$, since each square bracket does individually. Thus x is feasible for $P_{\mathscr{C}}(b)$, so that b is attainable and belongs to $\mathscr{B}$. That is, $\mathscr{B}$ is convex.

Suppose now that x', x'' solve $P_{\mathscr{C}}(b'), P_{\mathscr{C}}(b'')$ respectively. Then, since x is feasible for $P_{\mathscr{C}}(b)$, we have

$$U(b) \geqslant f(x) \geqslant pf(x') + qf(x'') = pU(b') + qU(b''),$$

which proves $U(b)$ concave.

It is remarkable that, in order to justify the strong Lagrangian principle for the case of equality constraints, one is forced to the very restrictive assumption that $g(x)$ is linear. Undoubtedly the conditions of Theorem (3.10) are unnecessarily strong, but it is not easy to find weaker conditions of a natural, general nature. Linearity of g has the simplifying feature that a supporting hyperplane to the graph of U at b is then immediately related to a supporting hyperplane to the graph of f at $x(b)$.

Concavity of U is the principal sufficient condition required in Theorem (3.9) for validity of the strong Lagrangian principle; is it also necessary? Essentially it is, if the principle is to hold for all b in $\mathscr{B}$, in that U must then coincide with its least concave majorant in $\mathscr{B}$; see Section (3.8).

It has been consistently assumed that x and $g(x)$ both take values in a finite dimensional space. However, applications often demand much more generality. For example, it may be necessary for x to take values in the space of functions (as in the continuous-time regulation problem of Section (6.3)) or in the space of distributions (see Section (5.4)). The methods used hitherto do indeed generalize naturally: convexity and the separating hyperplane theorem are fundamental notions. Nevertheless, treatment of more general cases does require additional discussion, concepts and precautions.

The interested reader is referred to the paper by Isii (1964), in which x is assumed to take values in a real vector space, $b - g(x)$ to take values in a linear topological space $\mathscr{C}$ and the form $y^{\mathrm{T}} z$ is replaced by the notion of a continuous linear functional $y(z)$ on $\mathscr{C}$.

Exercises

(3.28) Show that, under the hypotheses of Theorem (3.10), all solutions of $P_{\mathscr{C}}(b)$ form a convex set.

3.7 DUALITY

It was seen from equations (3.1) and (3.2) that, if a vector y exists such that $f(x) - y^{\mathrm{T}} g(x)$ is maximal at a value $\bar{x}$ which is feasible for $P_0(b)$, then $\bar{x}$

solves $P_0(b)$, and equality holds in the relation valid for all y:

$$U(b) \leqslant \max_x L(x, y). \qquad (3.55)$$

Consequently, this value of y must minimize the quantity

$$\max_x L(x, y) = \max_x [f(x) + y^T(b - g(x))] = \max_{\mathcal{B}} [U(\xi) + y^T(b - \xi)], \qquad (3.56)$$

and the minimal value attained is just $U(b)$.

In other words, if an appropriate value of the Lagrangian multiplier y exists, then it solves a minimization problem, the minimization of expression (3.56), and the minimum attained, $U(b)$, is just the value of the maximum attained for the original constrained maximization problem. This associated minimization problem is often termed the *dual problem* to $P_0(b)$, and $P_0(b)$ itself referred to as the *primal problem*.

The dual problem often has an interesting interpretation, particularly when the "price" characterization of y is recalled. For example, consider again the problem of Exercise (1.11): the determination of the flow of electric current is a resistive network. The problem will be framed slightly differently: let x_{jk} denote the current flowing from node j to node k of the network, so that the dissipation in the network is proportional to $\frac{1}{2} \sum_1^m \sum_1^m R_{jk} x_{jk}^2$. However, a current balance has to be observed at each node in the network, so that the problem is: the maximization of

$$f(x) = -\frac{1}{2} \sum \sum R_{jk} x_{jk}^2 \qquad (3.57)$$

subject to

$$\sum_k x_{jk} = b_j \quad (j = 1, 2, \ldots, m), \qquad (3.58)$$

where b_j is the net current fed externally into node j. Let multipliers y_j be associated with constraints (3.58), so that the Lagrangian form is

$$L(x, y) = -\frac{1}{2} \sum_j \sum_k R_{jk} x_{jk}^2 + \sum_j y_j \left(b_j - \sum_k x_{jk} \right). \qquad (3.59)$$

Maximizing freely, one finds the currents x_{jk} to be given, in terms of the multipliers y_j, by

$$x_{jk} = \frac{y_j - y_k}{R_{jk}} \qquad (3.60)$$

and that

$$\max_x L(x, y) = \frac{1}{2} \sum_j \sum_k \frac{(y_j - y_k)^2}{R_{jk}} + \sum_j b_j y_j. \qquad (3.61)$$

The classic interpretation of the y variables is clear from (3.50): y_j represents the the electrical potential at node j, and (3.60) is just a statement of Ohm's law. The relation $y = U_b{}^T$ gives another characterization of the potential y_j: as being proportional to the rate of change of total energy dissipation with current fed into node j.

The values of the potentials y_j solve the dual problem: they minimize expression (3.61). If potentials are considered only for those nodes which have no connections external to the network, and so for which $b_j = 0$, it is seen that these must minimize

$$D = \tfrac{1}{2} \sum_j \sum_k \frac{(y_j - y_k)^2}{R_{jk}}, \qquad (3.62)$$

so the dual gives the earlier form of the minimum dissipation principle taken in Exercise (1.11). The remaining potentials must either be determined by minimization of the full expression (3.61), or, in many cases, it is more realistic to suppose that it is the potential y_j which is prescribed at a node with external connections, rather than the current inflow b_j. That is, the coupling with the external world is made by the specification of "prices" rather than of flow rates. In such a case, then, those potentials which are not prescribed are determined by minimization of D.

It can, of course, be asked whether the minimization of expression (3.61) is an easier matter than the maximization of expression (3.57) subject to (3.58), and so whether the introduction of Lagrangian multipliers has effectively reduced the problem. If the number of conducting arcs (those for which $R_{jk} > 0$) is substantially greater than the number of nodes, then there is indeed an effective reduction. Moreover, the passage from a constrained problem to an unconstrained one is almost always worth while. For the continuum version of this problem, the expression of currents in terms of potentials is certainly worth while, since the current is a function of two arguments (the co-ordinates of two nodes) while the potential is a function of one only (the co-ordinate of a single node).

In general, the multipliers corresponding to constraints which express a balance or conservation condition will always have the character of a potential, with flow between points being determined by the potential difference, as in (3.60); see Sections (5.3) and (7.2). The linearity of the flow/potential-difference relation in the present case is a consequence of the quadratic dependence of energy dissipation-rate upon current.

For another example, consider the energy distribution problem of Exercise (2.5). If multipliers y_1 and y_2 are introduced to cope with the two equality constraints, then the Lagrangian form (3.3) is maximized by the Gibbs distribution (3.4). The multipliers satisfy the dual problem: they minimize

$$\sum_k e^{-y_1 - y_2 \varepsilon k} + N y_1 + \mathscr{E} y_2,$$

and the minimal value will equal the maximand of the primal: the entropy. The minimizing value of y_1 is easily seen to be

$$y_1 = \log \frac{1}{N} \sum_k e^{-y_2 \varepsilon k}$$

so that the entropy is given by

$$S = \text{const} + \min_{y_2} \left[\mathscr{E} y_2 + N \log \left(\sum_k e^{-y_2 \varepsilon k} \right) \right].$$

From the interpretation of the multipliers deduced at the end of Section (2.2) it is seen that this amounts to the thermodynamic identity

$$\text{entropy} = \frac{\text{energy} - \text{free energy}}{\text{temperature}}$$

with the additional assertion that, for a given energy, the temperature (proportional to y_2^{-1}) is such as to minimize this quantity. The entropy can thus be regarded as the extreme attained in both a maximization and a minimization problem: for other examples of "complementary variational principles" in physics, see Sections (6.5) and (10.4).

The duality principle for $P(b)$ demonstrated at the beginning of the section can be extended immediately to $P_\mathscr{C}(b)$ by formulating this problem as one with equality constraints, (2.46).

Theorem (3.11)

Suppose a vector y exists such that the value of (x, z) maximizing $f(x) + y^T(b - g(x) - z)$ in $(\mathscr{X}, \mathscr{C})$, say $(\bar{x}, \bar{z})$, is feasible for $P_\mathscr{C}(b)$. Then $(\bar{x}, \bar{z})$ solves $P_\mathscr{C}(b)$, and y minimizes.

$$\max_{\xi \in \mathscr{B}} [U(\xi) + y^T(b - \xi)] = \max_{\substack{x \in \mathscr{X} \\ z \in \mathscr{C}}} [f(x) + y^T(b - g(x) - z)],$$

the minimum equalling $U(b)$.

If $\mathscr{C}$ is a convex cone, then y minimizes $\max_{x \in \mathscr{X}} [f(x) + y^T(b - g(x))]$ in $\mathscr{C}^*$, the minimum equalling $U(b)$.

The first part of the theorem is proven. To see the reduction expressed in the second part, note that, for any y in $\mathscr{C}^*$,

$$U(b) = \max_{\mathscr{X}(b)} f(x) \leqslant \max_{\mathscr{X}(b)} [f(x) + y^T(b - g(x))]$$

$$\leqslant \max_x [f(x) + y^T(b - g(x))]. \tag{3.63}$$

Now, a vector y such that

$$f(\bar{x}) - y^{\mathrm{T}}(g(\bar{x}) + \bar{z}) \geqslant f(x) - y^{\mathrm{T}}(g(x) + z) \tag{3.64}$$

for (x, z) in $(\mathscr{X}, \mathscr{C})$ must belong to $\mathscr{C}^*$ and satisfy

$$y^{\mathrm{T}} \bar{z} = y^{\mathrm{T}}(b - g(\bar{x})) = 0, \tag{3.65}$$

as is known from Theorem (2.7). Setting this value of y in the last member of (3.63) one sees that the maximum is attained at $\bar{x}$, and is equal to

$$f(\bar{x}) + y^{\mathrm{T}}(b - g(\bar{x})) = f(\bar{x}) = U(b).$$

Equality thus holds in (3.63), so that the multiplier y with property (3.64) does minimize the last member of (3.63) in $\mathscr{C}^*$, as asserted.

So, for the allocation problem, when $\mathscr{C} = R_+^m$, then y minimizes

$$\max_{\mathscr{X}} [f(x) + y^{\mathrm{T}}(b - g(x))] \tag{3.66}$$

subject to $y \geqslant 0$, and it satisfies (3.65). It is known that y has a price interpretation in this case, either as $y = U_b^{\mathrm{T}}$, or in some more general sense U_b does not exist. The bracket in (3.66) thus represents the combined value of return on activities $f(x)$ and unused resources $y^{\mathrm{T}}(b - g(x))$. The optimizer chooses his allocation x freely in $\mathscr{X}$ to maximize this total value. Prices y are then fixed to minimize this optimized total value, subject to the restriction that prices should be non-negative. These minimizing prices will attach zero value to unused resources.

The discussion at the end of Section (3.1) gave a geometric characterization of the dual problem, which will be extended in the next section. The relation between primal and dual can be also seen as a limit case of the relation between an integral transform and its inverse; see Sections (8.6) and (8.7).

Other dualizations of a problem are possible (see Roode, 1968) based essentially on the notion of representing the graph of $U(b)$ as the envelope of a family of surfaces other than hyperplanes. However, the conventional theory, based on the notion of supporting hyperplanes to this graph, has a special status. In it, the multipliers y have the gradient or price interpretation ($y = U_b^{\mathrm{T}}$, in some sense) and the minimum for the dual, $\hat{U}(b)$, represents the maximum that can be achieved in the primal by use of randomized solutions.

Exercises

(3.29) Show that an appropriate multiplier vector y does exist for the two problems discussed in the text (current flow, and energy distribution).

(3.30) Consider the maximization of a concave quadratic function $f(x)$ in $\mathscr{X}$, subject to the single linear constraint $b - a^{\mathrm{T}} x \in \mathscr{C}$, in the various combinations of cases when (i) $\mathscr{X}$ is the real axis, or just the positive real axis, (ii) $\mathscr{C}$ is the positive axis, or just the origin. Determine the dual problem in each case and verify that it yields the same extreme value as the primal.

3.8 LAGRANGIAN METHODS IN THE NON-CONVEX CASE; RANDOMIZED SOLUTIONS

We wish now to generalize the graphical argument associated with Figure (3.2), and show that the bound

$$\hat{U}(b) = \min_{y} \max_{\xi \in \mathscr{B}} [U(b) + y^{\mathrm{T}}(b - \xi)] \tag{3.67}$$

is one that can be attained if randomized solutions are admitted. For simplicity the argument will be restricted to the case of equality constraints, $P(b)$. It transfers immediately to the only apparently more general case $P_{\mathscr{C}}(b)$, especially since much of the argument is directly in terms of U and $\mathscr{B}$.

The smallest closed convex set containing $\mathscr{B} = g(\mathscr{X}]$ is the *closed convex hull* of $\mathscr{B}$, denoted $[\mathscr{B}]$. Consider now a class $\mathscr{E}$ of expectation operators E on $\mathscr{X}$, which contains the operators corresponding to all the one-point distributions on $\mathscr{X}$, and is closed under averaging, i.e. $E, E' \in \mathscr{E}$ implies that $pE + qE' \in \mathscr{E}$. Then the set of points representable as $Eg(x)$ for some E in $\mathscr{E}$ is also a point set in R^m; let its closure be denoted by $\{\mathscr{B}\}$.

Theorem (3.12)

$$\{\mathscr{B}\} = [\mathscr{B}].$$

From the defining properties of $\mathscr{E}$ it follows that $\{\mathscr{B}\}$ is convex and contains $\mathscr{B}$; consequently $\{\mathscr{B}\} \supset [\mathscr{B}]$. On the other hand, if $\{\mathscr{B}\}$ contained a point ζ not in $[\mathscr{B}]$, then there would exist a hyperplane strictly separating ζ and $[\mathscr{B}]$, i.e. a coefficient vector α such that

$$\alpha^{\mathrm{T}} \zeta > \alpha^{\mathrm{T}} \xi$$

for all ξ in $[\mathscr{B}]$, and so for all ξ in $\mathscr{B}$. That is,

$$E(\alpha^{\mathrm{T}} g(x)) > \alpha^{\mathrm{T}} g(x) \tag{3.68}$$

for all x in $\mathscr{X}$, and for E the particular expectation operator on $\mathscr{X}$ corresponding to ζ. But the strict inequality (3.68) is inconsistent with the properties of an expectation. Hence all points ζ of $\{\mathscr{B}\}$ belong to $[\mathscr{B}]$, and the assertion of the theorem follows.

In fact, it is known from Exercise (3.15) that, if $\mathscr{B}$ is bounded, then any point of $[\mathscr{B}]$ can be represented as an average of at most $m + 1$ points of $\mathscr{B}$, and so can be represented $Eg(x)$, where the operator E corresponds to a distribution with mass on at most $m + 1$ points of $\mathscr{X}$.

Consider now the set $\mathscr{D} = (g(\mathscr{X}), f(\mathscr{X}))$ in R^{m+1}, supposed closed, whose upper boundary has the equation

$$\eta = U(\xi) \quad (\xi \in \mathscr{B}) \tag{3.69}$$

That is, $U(\xi)$ is the greatest value of η consistent with $(\xi, \eta) \in \mathscr{D}$. Suppose that the convex hull of $\mathscr{D}$, $[\mathscr{D}]$, has correspondingly an upper boundary with equation

$$\eta = \breve{U}(\xi). \tag{3.70}$$

Then $\breve{U}$ is a function concave in $[\mathscr{B}]$, and is, in fact, the least concave majorant of U on $\mathscr{B}$.

It follows from Theorem (3.12) that $[\mathscr{D}] = \{\mathscr{D}\}$, so that $\breve{U}(b)$ can be interpreted as the greatest value of η consistent with $(b, \eta) \in \{\mathscr{D}\}$, or as the supremum of $Ef(x)$ over all expectation operators E in $\mathscr{E}$ consistent with

$$Eg(x) = b. \tag{3.71}$$

That is, $\breve{U}(b)$ is the analogue of $U(b)$ for $P(b)$ if randomized solutions are permitted.

Theorem (3.13)

Suppose that $\breve{U}(b)$ obeys any of the conditions (a)–(c) of Theorem (3.9) (iii). Then

$$\breve{U}(b) = \hat{U}(b). \tag{3.72}$$

Furthermore, if a supporting hyperplane to $[\mathscr{D}]$ at $(b, \breve{U}(b))$ meets $[\mathscr{D}]$ only in a bounded point set, then the bound $\hat{U}(b)$ can be attained by a randomized solution on at most $m + 1$ points of $\mathscr{X}$.

That is, under mild conditions the bound $\hat{U}(b)$ is the supremum of values $Ef(x)$ attainable using distributions feasible for $P(b)$; under further mild conditions this supremum is itself attainable, and by a randomized solution involving at most $m + 1$ x-values.

Let $\mathscr{E}(b)$ denote the set of elements of $\mathscr{E}$ consistent with (3.71); the expectation operators "feasible" for $P(b)$. Then for any y

$$\breve{U}(b) = \sup_{\mathscr{E}(b)} Ef(x) = \sup_{\mathscr{E}(b)} E[f(x) + y^{\mathrm{T}}(b - g(x))]$$

$$\leqslant \max_{x} [f(x) + y^{\mathrm{T}}(b - g(x))].$$

Minimizing the right-hand member with respect to y, one sees that

$$\breve{U}(b) \leqslant \hat{U}(b). \tag{3.73}$$

On the other hand, since $(b, \breve{U}(b))$ is a boundary point of the convex set $[\mathscr{D}]_+$, it follows that there exists a vector y and a non-negative scalar w

such that $w\eta - y^{\mathrm{T}}\xi$ is maximal for (ξ, η) in $\mathscr{D}$ at $(b, \breve{U}(b))$ and, in particular, such that

$$w\breve{U}(b) - y^{\mathrm{T}}b \geqslant w\breve{U}(\xi) - y^{\mathrm{T}}\xi \quad (\xi \in \mathscr{B}). \qquad (3.74)$$

Furthermore, the initial assumption of Theorem (3.13) allows w to be normalized to unity. It follows then from (3.74) that

$$\breve{U}(b) \geqslant \max_{\xi} [\breve{U}(\xi) + y^{\mathrm{T}}(b - \xi)]$$

$$\geqslant \max_{\xi} [U(\xi) + y^{\mathrm{T}}(b - \xi)]$$

$$\geqslant \hat{U}(b) \qquad (3.75)$$

and, from (3.73) and (3.75), assertion (3.72) follows.

To prove the second assertion, note first that at least one supporting hyperplane exists, since $[\mathscr{D}]$ is convex. Denote the point set it constitutes by $\mathscr{S}$. Then $(b, \breve{U}(b)) = (b, \hat{U}(b))$ belongs to the m-dimensional point set $\mathscr{S} \cap [\mathscr{D}] = [\mathscr{S} \cap \mathscr{D}]$ (see Exercise (3.14)), and so, by Exercise (3.15), is representable as an average of at most $m+1$ points of $\mathscr{S} \cap \mathscr{D}$.

A case where concavity of $U(b)$ fails is that where the attainable set $\mathscr{B}$ is non-convex. For a simple example, consider the case of a retailer maximizing $f(x)$ subject to $x = b$ (on average, at least), where $f(x)$ is his return from

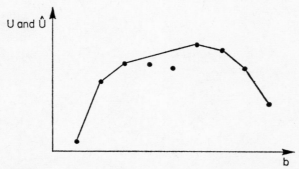

Figure 3.8 Functions U and $\hat{U}$ for a case where $\mathscr{B}$ consists only of the non-negative integers, and so is non-convex. Values of U are marked with heavy dots; values of $\hat{U}$ by the continuous line

selling an amount x of a scalar commodity, of which there are supplies b. Suppose that the commodity is handled only in multiples of a basic unit, so that x can only take the values $0, 1, 2, \ldots$. Then both $\mathscr{X}$ and $\mathscr{B}$ are the set of integers, a discrete and non-convex set, and to prescribe any but an integral value for b is meaningless. $U(b)$ equals $f(b)$ for b integral; elsewhere it is undefined. However, if randomized strategies are admitted, then $\mathscr{X}$ and $\mathscr{B}$ are both extended to R_+^1, and the optimal return then attainable, $\hat{U}(b)$, is defined for all non-negative b. The value of $\hat{U}(b)$ for non-integral b will be a linear interpolation between values of $U(b)$ ($=f(b)$, in this case), for integral b; this can be attained only by randomization. For integral b, one will have $U(b) = \hat{U}(b)$ iff $U(b)$ coincides with its concave majorant at this point, when $\hat{U}(b)$ can be attained by a one-point solution.

Exercise

(3.31) Consider a vector x in R^n of length $|x|$. Maximize $E[|x|e^{-|x|}]$ over distributions on R^n consistent with $E(x) = 0$. What are the extremal distributions?

CHAPTER 4

Linear Programming

Linear programming is certainly the best known and most widely used of the optimization techniques developed in recent years. Many practical problems present themselves in the form: maximization of a linear function in, say, the positive orthant, subject to linear constraints. This structure gives the problem a very attractive and symmetric Lagrangian theory. Furthermore, problems amenable to strong Lagrangian methods can, in a certain sense, all be represented as linear programmes: see Section (8.3).

Ultimately one must resort to numerical methods. Of these, Dantzig's simplex method and its variations are the classic ones, with a strong basis in Lagrangian theory.

4.1 CHARACTERIZATION OF A LINEAR PROGRAMMING PROBLEM

Suppose that the constrained maximization problem defined in Section (2.6) has the following special features: f and g are linear; $\mathcal{X}$ and $\mathcal{C}$ are cones. Then it is termed a *linear programming* (abbreviated LP) problem. Such a problem has special features, and special techniques and results are available for it.

Note, first, that the linearity of f implies that the solution $x(b)$ will in general be on the boundary of the feasible region $\mathcal{X}(b)$, and may always be chosen on the boundary. For a linear function has a stationary point in a region if and only if it is constant in that region (see Exercises (4.1) and (4.2)).

One of the most typical linear programming situations is the allocation problem: the linear version of the problem described in Section (2.1). It is desired to maximize

$$f(x) = c^{\mathrm{T}} x \tag{4.1}$$

71

subject to

$$x \geqslant 0, \tag{4.2}$$

$$Ax \leqslant b. \tag{4.3}$$

The interpretation of the intensities x_k and of the inequalities (4.2) and (4.3) is as before: intensities are intrinsically non-negative, and resources consumed must not exceed resources available. The element c_k of c represents the return to be made on activity k per unit intensity. The linear form of f implies that returns from different activities are proportional to scale, and also additive; there is no interaction between activities. Amounts of a resource consumed by different activities are likewise proportional to scale and additive; a_{jk} represents the amount of resource j absorbed by activity k per unit intensity.

A geometric view of the situation is helpful: a two-dimensional situation is presented in Figure (4.1). The region $\mathscr{X}(b)$ determined by (4.2) and (4.3) will

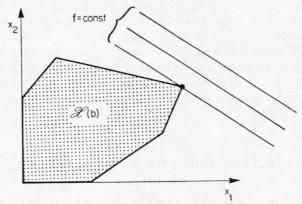

Figure 4.1 The feasible set $\mathscr{X}(b)$ for the allocation problem is a convex polyhedron. The point in this set maximizing $f = c^{\mathrm{T}} x$ will lie on its boundary, and can always be chosen on a vertex

be the intersection of a number of half-spaces and hence will be a convex polyhedron in the positive orthant of x-space: see Exercise (3.7). If we now consider the family of hyperplanes

$$f \equiv c^{\mathrm{T}} x = \text{const.} \tag{4.4}$$

then the aim is to decrease the constant until the hyperplane (4.4) first has a point in common with $\mathscr{X}(b)$. The common point is $x(b)$, and the value of the constant is $U(b)$. The hyperplane (4.4) will indeed meet $\mathscr{X}(b)$ in a boundary point, and, in fact, usually in a vertex, as is plain from the diagram. The only case when $x(b)$ is not necessarily on a vertex is when the hyperplane meets a whole face (of some dimensionality) of $\mathscr{X}(b)$. Then all points in the face are solutions. However, these include a number of vertices, the extreme points of the face, so that there is always a vertex of $\mathscr{X}(b)$ which provides a solution. This assertion is proved formally later in Theorem (4.2).

One can see that there are two situations which are in some sense anomalous. One is that in which relations (4.2) and (4.3) are mutually inconsistent, so that $\mathscr{X}(b)$ is empty, and the problem simply has no solution. Another situation is that in which (4.2) and (4.3) allow some x_k to take infinite values, as in Figure (4.2), and x can be varied in such a direction as

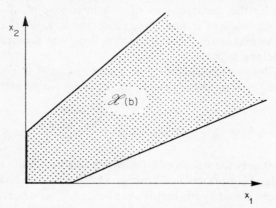

Figure 4.2 A problem with a feasible set extending to infinity

to make $c^T x$ indefinitely large. In this case $P(b)$ has a solution, but with $U(b) = +\infty$. However, a real-life problem should not present either of these features: if it does, it has been improperly formulated. For example, it is found in Section (5.1) that an attempt to find an optimum equilibrium economy leads to $U = +\infty$ if the economy is, in fact, capable of indefinite growth.

For an allocation problem, the obvious physical condition that precludes infinite solutions in the model considered hitherto is that all activities require some resources in order to operate, but themselves produce no resources. That is, $a_{jk} \geq 0$ with strict inequality for some j for every k, so that then

$$x_k \leqslant \min_j \left(\frac{b_j}{a_{jk}} \right) < \infty.$$

If $g(x) = Ax$ then the attainable set of x-values could be written

$$\mathscr{B} = A\mathscr{X} + \mathscr{C}.$$

Note that, since $\mathscr{X}$ and $\mathscr{C}$ are both cones, then so is $\mathscr{B}$; if they are both convex, so is $\mathscr{B}$; if they are both polyhedral, so is $\mathscr{B}$.

Exercises

(4.1) Note that if a linear function $c^T x$ is maximal with respect to variations $x \to x + \varepsilon s$ (s being a fixed direction and ε taking at least one positive value) then $c^T s \leqslant 0$. Hence show that $c^T x$ always attains its maximal value in $\mathscr{X}$ on the boundary of $\mathscr{X}$, and attains it in the interior only if it is constant.

(4.2) Consider the maximization of $c^T x$ in a cone $\mathscr{X}$. Show that, if the maximum is attained at a non-zero, finite value $\bar{x}$, then $c^T \bar{x} = 0$. Show further that the maximum value attained is 0 or $+\infty$ according as $-c$ does or does not belong to $\mathscr{X}^*$.

(4.3) Consider the maximization of $\sum c_k x_k$ in $x \geqslant 0$ subject to the single constraint $\sum a_k x_k = b$. Show that, if the c_k are positive and the a_k strictly positive, then

$$U(b) = \max_k \left(\frac{bc_k}{a_k} \right),$$

and that $x_k > 0$ only for values of k for which this maximum is attained.

The horse-backing example of Exercise (1.10) would reduce to this case if the total amount stated by the bettor were limited, $\sum x_k \leqslant b$, and the value of b so small that $f(x)$ could be approximated by the linear function

$$\alpha(\textstyle\sum t_k)(\sum p_k x_k / t_k) - \sum x_k$$

Compare the solution of this problem with that of Exercise (1.10).

4.2 LAGRANGIAN METHODS; THE DUAL PROBLEM

It may be supposed that the linear functions $f(x)$ and $g(x)$ can be written in the homogeneous forms $c^T x$ and Ax respectively: any constant term can be removed by a redefinition of f or of b.

If the cones $\mathscr{X}$ and $\mathscr{C}$ are convex then the sets

$$\mathscr{B} = A\mathscr{X} + \mathscr{C}, \quad \text{and} \quad \mathscr{D} = (\mathscr{B}, f(\mathscr{X}))$$

are also convex, so that $U(b)$ is concave in $\mathscr{B}$. The concavity assumptions of Theorem (3.9) are therefore satisfied, and the theorem can be reformulated for the special case of a linear programme as follows.

Theorem (4.1)

Suppose $\bar{x}$ solves the problem $P_{\mathscr{C}}(b)$ of maximizing $f(x) = c^T x$ in a cone $\mathscr{X}$ subject to the constraint that $b - Ax$ lies in a cone $\mathscr{C}$. Suppose $\mathscr{C}, \mathscr{X}$ convex, and $b, \bar{x}$ finite.

(i) A vector y and a non-negative scalar w exist such that $\bar{x}$ maximizes $wc^T x - y^T Ax$ in $\mathscr{X}$.

(ii) The directional derivatives of U at b satisfy

$$y^T s \geqslant w D_s U$$

for all directions s from b in $\mathscr{B}$.

(iii) The constant w may be normalized to unity if either (a) b is interior to $\mathscr{B}$ and U finite in a neighbourhood $N(b)$ of b or (b) $\mathscr{X}$ and $\mathscr{C}$ are polyhedral and U finite in $N(b) \cap \mathscr{B}$.

(iv) If b is interior to $\mathscr{B}$ and U_b exists and is finite, then w may be normalized to unity, and y necessarily then equals U_b^T.

(v) If w can be normalized to unity then the vector y minimizes $y^T b$ in $\mathscr{C}^*$ subject to $A^T y - c \in \mathscr{X}^*$. This is the problem dual to $P_{\mathscr{C}}(b)$, and the minimum attained equals $U(b)$.

(vi) y and $\bar{x}$ satisfy

$$y^T(b - A\bar{x}) = 0, \tag{4.5}$$

$$(c^T - y^T A)\bar{x} = 0. \tag{4.6}$$

Assertions (i)–(iv) transfer directly from Theorem (3.9); the proof of (iii) appeals to the fact that $\mathscr{D}$ is polyhedral ($U(\xi)$ is piecewise linear) if $\mathscr{X}$ and $\mathscr{C}$ are polyhedral.

The other statements stem from the specifically LP nature of the problem. Suppose that a linear form $d^T x$ is being maximized in a cone $\mathscr{X}$. The basic variation available is $x \to \lambda x$, where λ is a non-negative scalar. If the maximum is attained at a finite value $\bar{x}$, we must thus have

$$d^T \bar{x} = 0. \tag{4.7}$$

If $d^T x \leqslant 0$ for all x in $\mathscr{X}$, then zero is the maximal value, certainly attainable at $x = 0$. However, if $d^T x > 0$ for some x in $\mathscr{X}$, then the maximum is $+\infty$, approached as x goes to infinity along a particular ray in $\mathscr{X}$. In summary,

$$\max_{x} [d^T x] = \begin{cases} 0, & -d \in \mathscr{X}^*, \\ +\infty, & \text{otherwise.} \end{cases} \tag{4.8}$$

Apply this argument now to the maximization of the form $c^T x - y^T Ax$. Since the maximum is attained at the finite value $\bar{x}$, (4.6) follows from (4.7). Furthermore

$$\max_{x} [c^T x - y^T Ax] = \begin{cases} 0, & \text{if } A^T y - c \in \mathscr{X}^*, \\ +\infty, & \text{otherwise,} \end{cases} \tag{4.9}$$

by (4.8).

Now, we know from Theorem (3.11) that the multiplier vector y satisfies (4.5), and minimizes $\max_x [c^T x + y^T(b - Ax)]$ subject to $y \in \mathscr{C}^*$. But it is seen from (4.9) that the only y values which could conceivably furnish a minimum are those for which $A^T y - c \in \mathscr{X}^*$, and that in this region the form minimized is just $y^T b$. Hence assertion (v).

The interesting point is the completely reciprocal character of the primal and dual problems: both are LP in nature, and the primal is the dual of the dual, as the reader can easily verify. Relation (4.6) is reciprocal in the same sense to relation (4.5).

Of course, this symmetry reveals a fact which may seem disturbing: that in solving an LP problem by Lagrangian methods one is simply converting it into another LP problem, in the dual variable y. There is thus no reduction of the problem unless m is substantially less than n, so that the number of variables in the dual problem is substantially less than in the primal. However, this would be true of the Lagrangian method generally.

The "price" interpretation of the dual variables often gives the dual problem itself an interesting interpretation, as will be seen in the next and later sections.

The properties of the dual problem mirror those of the primal. For example, *if the primal problem has an infinite solution, $U = +\infty$, then the dual problem has no solution*. For we know that, for any y in $\mathscr{C}^*$,

$$U(b) \leqslant \max_x [c^T x + y^T(b - Ax)], \tag{4.10}$$

see (3.63). Since $U = +\infty$, the right-hand member of (4.10) must be unbounded above. Thus, for any y in $\mathscr{C}^*$ there is an x in $\mathscr{X}$ for which $c^T x - y^T Ax > 0$, so that $A^T y - c$ cannot belong to $\mathscr{X}^*$. Consequently, there is no y feasible for the dual. One can complete this result in various ways.

Exercises

(4.4) Suppose that feasible finite values exist for both primal and dual. Show that both then have finite solutions.

(4.5) What is the problem dual to the first example of Exercise (4.3)? Verify the equality of the two extreme values. Consider the form of the dual in cases (i) where, for some k, $a_k = 0$ and $c_k > 0$, so that $U = +\infty$; (ii) where $a > 0$ and $b < 0$, so that the primal has no solution.

4.3 ALLOCATION; INTERPRETATION OF THE DUAL

Consider again the allocation problem described by relations (4.1)–(4.3). This also goes under the name of the *activity analysis* problem, for obvious reasons.

Among LP problems, this one is characterized by the fact that $\mathscr{X}, \mathscr{C}$ are the positive orthants in their respective spaces. The same will then be true of $\mathscr{X}^*, \mathscr{C}^*$ so that the dual form of the problem is: minimize

$$F(y) = y^{\mathrm{T}} b, \tag{4.11}$$

subject to

$$y \geqslant 0, \tag{4.12}$$

$$y^{\mathrm{T}} A \geqslant c^{\mathrm{T}}. \tag{4.13}$$

Since the set $\mathscr{D}$ is polyhedral, then U is piecewise linear, and the derivative U_b exists almost everywhere. It is known that $y = U_b{}^{\mathrm{T}}$ where this exists, so that y can then be interpreted as a vector of effective marginal prices or *shadow prices* for the various resources. That is, an increase by δ in the amount of resource j makes possible an increase in optimal return of $y_j \delta + o(\delta)$.

The value of U_b is ambiguous at points where U changes from one linear form to another. These are the points at which a different set of constraints becomes effective, and the basis (see Section (4.5)) of the solution changes. In this case y can still be given a modified price interpretation, because of its relation to the directional derivatives of U (see Exercise (4.6)). The implications of relation (4.5) in the case $\mathscr{C} = R_+{}^m$ have already been mentioned:

$$y_j = 0 \quad \text{if} \quad (A\bar{x})_j < b_j, \tag{4.14}$$

and

$$(A\bar{x})_j = b_j \quad \text{if} \quad y_j > 0. \tag{4.15}$$

That is, a resource which is not fully utilized in the optimal solution has zero marginal price, and a resource with positive marginal price is fully utilized. This is reasonable; one would pay a positive amount for additional supplies only if existing supplies were being fully used.

Correspondingly, it follows from relation (4.6), the relation dual to (4.5), that

$$\bar{x}_k = 0 \quad \text{if} \quad (y^{\mathrm{T}} A)_k > c_k, \tag{4.16}$$

and

$$(y^{\mathrm{T}} A)_k = c_k \quad \text{if} \quad \bar{x}_k > 0. \tag{4.17}$$

Now, $(y^{\mathrm{T}} A)_k$ is the value (on shadow prices) of resources used in running activity k at unit intensity, so that $c_k - (y^{\mathrm{T}} A)_k$ is the profit to be made from this activity per unit intensity. These relations then state that an unprofitable activity is not undertaken, and an activity which is undertaken just breaks even.

At the end of Section (3.7) a characterization was given of the dual of an allocation problem. For the linear problem, whose dual is expressed in relations (4.11)–(4.13), this characterization can be sharpened: *shadow prices adopt those values (non-negative) which minimize the value of resources, subject to the condition that no activity shows a positive profit.*

It is interesting that the original maximization problem, which seemed to involve only one party (the decision-maker, or entrepreneur) brings in, through the dual problem, the concept of an effective "shadow opponent", whose perverse aim it is to fix effective prices at the values which are most unfavourable to the decision-maker. The constraints under which the decision-maker is acting can be construed as the effect of the existence of such a mock opponent.

Of course, the fact that an activity makes zero profit does not mean that it is "unprofitable" in the conventional sense. The entrepreneur has realized a return by using the resources available to him (which may include his own time and energy) and the statement of "zero profit" is merely a statement of the fact that he has received a fair return for them.

The combined primal/dual problem can be stated as the finding of an x and a y to yield the double extremal

$$U(b) = \min_{y \geqslant 0} \max_{x \geqslant 0} [c^T x + y^T(b - Ax)]. \tag{4.18}$$

The decision-maker and his shadow opponent fix allocation and prices, respectively, in such a way as to respectively maximize and minimize total value of activities and of unused resources.

The description (4.18) of the problem as a simultaneous maximization and minimization bring it near indeed to formulation as a *game* (see Sections (10.4), (10.5)) with two players, of conflicting interests. The Lagrangian assertions of Theorem (4.1) amount to the statement that the minimization and maximization operations of (4.18) commute, so that the order of the two operations is immaterial; see Exercise (3.1). This is certainly a requirement, if the concept of a game with simultaneous moves by the two players is to have any meaning.

Exercises

(4.6) If $U(b)$ is not differentiable at a point, then the effective marginal price depends upon the direction in which one varies b. Suppose that one envisages an increase in b_j alone. Show that the effective marginal price for resource j is then the *least* value of y_j consistent with y's being an appropriate multiplier value for the problem (cf. Exercise (3.27)). What would be the analogous assertion if a decrease in b_j alone were considered?

4.4 SOME OTHER LINEAR PROBLEMS

(a) The diet problem

The aim is to construct a minimum cost diet, using foodstuffs available, which does not fall below certain nutritional requirements. Suppose that foodstuff j costs b_j per unit quantity, and contains an amount a_{jk} of nutrient k per unit quantity. The total amount of nutrient k in the diet must not fall below c_k.

The problem is then to minimize $y^T b$ subject to $y \geqslant 0$ and $y^T A \geqslant c^T$; exactly the dual of the allocation problem, as the notation we have chosen emphasizes.

(b) The transportation problem

This is one of the classic LP situations, with a large literature of its own. The aim is to transport a single bulk commodity (say, oil) from the basic supply points (depots) to the consumption points (destinations) in a minimum-cost fashion. The amounts transported must be such that demand is satisfied at each destination, and supply not exceeded at each depot.

So, suppose that x_{jk} is the amount transported from depot j to destination k, at a cost c_{jk} per unit amount $(j = 1, 2, ..., m; k = 1, 2, ..., n)$. It is thus required to minimize

$$f(x) = \sum_j \sum_k c_{jk} x_{jk}, \tag{4.19}$$

subject to

$$x_{jk} \geqslant 0 \tag{4.20}$$

and

$$\sum_k x_{jk} \leqslant r_j, \tag{4.21}$$

$$\sum_j x_{jk} \geqslant s_k. \tag{4.22}$$

Here r_j is the amount available at depot j, and s_k the amount required at destination k.

Because the problem has such a particular structure it has its own individual and rather attractive algorithm, described in Section (4.8). The problem can also be regarded as a special case of the large class of "flow" problems, concerned with optimization of the flow of one or more commodities through a given network of routes.

(c) Routing problems as networks

This, like the transportation problem, is an example of a flow problem. Suppose that one has a network of telegraph stations, joined by links of given capacities. One wishes to route telegrams through the network, paying regard to limitations of capacity, etc., in such a way as to optimize the overall running of the system, in some sense.

Let the stations of the network be referred to as *nodes*. Suppose that traffic for destination i originates at node j at rate r_{ij} (per unit time). It will be assumed that the system is in equilibrium, and the traffic is routed in such a way that traffic for destination i is transmitted along the jk link (in the direction $j \to k$) at rate x_{ijk}. There will then be a balance relation for traffic through node j:

$$r_{ij} = \sum_k (x_{ijk} - x_{ikj}) \quad (i \neq j). \tag{4.23}$$

This is to be supplemented by the natural relations

$$\left. \begin{array}{l} x_{iik} = 0, \\ x_{ijj} = 0. \end{array} \right\} \tag{4.24}$$

If the jk link has capacity b_{jk}, then there will be the further restriction

$$\sum_i x_{ijk} \leqslant b_{jk}. \tag{4.25}$$

The problem is now to minimize

$$f(x) = \sum_i \sum_j \sum_k c_{ijk} x_{ijk} \tag{4.26}$$

with respect to the x_{ijk}, subject to $x_{ijk} \geqslant 0$ and constraints (4.23)–(4.25). Here c_{ijk} is some measure of the direct cost of sending i-destination traffic along the jk-link; this may well be independent of i. For example, c_{ijk} might be some measure of transmission-time or transmission-distance along the jk-link, in which case average total time or distance taken for all traffic is being minimized.

Here c_{ijk} was referred to as a "direct cost". This is because the dual formulation will throw up the notion of a "shadow cost", which is economically the meaningful quantity. This shadow cost will take account, not merely of direct charges, but also of losses due to the capacity restrictions (4.25). The shadow costs will operate in such a way as to ease traffic off congested links.

Of course, one objection to a criterion function such as (4.26) is that it corresponds merely to an overall, average evaluation. It may lead to the policy that one routes some small proportion of traffic in a circuitous and time-consuming fashion, in order to be able to route the bulk of the traffic more directly. A criterion sensitive to the rights of minorities would penalize such a policy, although at the expense of adopting a nonlinear criterion function.

Another problem is that in which the "traffic" consists of say, commodities, rather than telegrams, and there are several types of commodity. Some of these may be destination-labelled (e.g. telegrams), some not (e.g. oil).

Section (5.3) is devoted to discussion of a particular flow problem in more detail.

(d) The warehousing problem

This is the case of a wholesaler who wishes to optimize his trading (in a single, divisible commodity) over a period. The period is supposed divided into convenient stages, say days. Suppose that during day k he sells an amount x_k at price b_k, and at the end of the day buys an amount w_k at price c_k. The sequences b_k and c_k are assumed to be known in advance.

If S_k is the warehouse stock at the beginning of the kth day, then

$$S_{k+1} = S_k - x_k + w_k, \tag{4.27}$$

and

$$S_{k+1} = S_0 + \sum_{j=1}^{k} (w_j - x_j). \tag{4.28}$$

The trader's profit over a period $k = 1, 2, ..., N$ is

$$f(x, w) = \sum_{1}^{N} (b_k x_k - c_k w_k), \tag{4.29}$$

and he wishes to choose the sequences x, w to maximize this, under the restrictions

$$x_k \geqslant 0, \quad w_k \geqslant 0, \tag{4.30}$$

$$x_{k+1} \leqslant S_0 + \sum_{1}^{k} (w_j - x_j) \leqslant C, \tag{4.31}$$

where C is the capacity of the warehouse. Inequalities (4.31) express the constraints that sales cannot exceed stock, and stock cannot exceed capacity.

The problem is a linear one, and can be treated by LP methods. However, its dynamic character makes a solution by dynamic programming more natural (see Exercise (4.9)); all the more so as this would be the only method available in the more realistic case of uncertain future prices.

Because of the assumptions of certainty and linearity, the policy generally takes (and always can take) an extreme form: the trader sells either all his stock or none of it; he buys either to capacity or not at all.

Exercises

(4.7) Formulate and interpret the problems dual to those described in the text.

(4.8) Show how the dual of the warehousing problem may be solved. Use Lagrangian methods to confirm the assertions made at the end of Section (4.4d).

(4.9) Let $V_k(S)$ be the maximal future profit that can be made in the warehousing situation, if the trader is at the beginning of the kth day with stock S. Show that $V_{N+1}(S) = 0$ and that

$$V_k(S) = \max_{0 \leqslant x \leqslant S} \max_{0 \leqslant w \leqslant c - S + x} [b_k x - c_k w + V_{k+1}(S - x + w)],$$

the inner maximization being taken first. Hence show that $V_k(S)$ is linear in S, and deduce again the assertions made at the end of Section (4.4d).

4.5 THE SIMPLEX METHOD: RATIONALE

The analytic treatment of Section (4.2) can be continued further only if the linear programming problem has some rather special structure; some such cases will be discussed in the next chapter. However, if the elements of A, b and c are arbitrary in value, then there is no recourse but that of numerical computation.

There are several computational methods for LP problems, some of them specially adapted to particular cases. However, the original and principal one is Dantzig's *simplex method*, which will now be described.

A preliminary to the simplex method is the transformation of the problem to a standard form with equality constraints: the maximization of $f(x) = c^T x$ subject to $x \geqslant 0$ and

$$Ax = b. \tag{4.32}$$

Not all linear problems are reducible to this form, but the allocation problem of Section (4.3) is, for example. In this case there is a set of inequalities

$$\sum_{k=1}^{n} a_{jk} x_k \leqslant b_j \quad (j = 1, 2, \ldots, m) \tag{4.33}$$

By the introduction of *slack variables* $x_{n+1}, \ldots, x_{n+m}$ these can be rewritten in the form

$$\sum_{k=1}^{n} a_{jk} x_k + x_{n+j} = b_j. \tag{4.34}$$

The additional variables are necessarily positive, and, with a redefinition of the vector x and the matrix A, we have achieved the equality form of the constraint, (4.32).

The vector of slack variables is, of course, just the vector z of Section (2.6), now incorporated explicitly into a redefined x vector. For the allocation problem it is to be interpreted as the vector of amounts of resources unused. By introduction of appropriate slack variables, all the problems of Section (4.4) can be reduced to the standard form.

The problem is now assumed posed in standard form with A an $m \times n$ matrix, so that m is the number of constraints, and n the number of variables. A feasible vector x for this problem is said to be *basic* if not more than m of its elements are non-zero. The set of k for which $x_k > 0$ is termed the *basis* of the vector.

Theorem (4.2)

The optimal solution may be taken as basic.

This is just a restatement of the geometrical conclusion of Section (4.1): that the solution can always be taken on a vertex of the polyhedron constituted by the feasible x-set. For the particular case of the allocation problem, the theorem states that, in the optimal solution, the sum of the numbers of activities actually pursued and of resources not fully utilized need not exceed the number of resource constraints.

The case in which A is of less than full rank (which it will be if $m > n$) can be excluded. In such a case, some of the constraints (4.32) are implied by the others: deleting these, the problem is reduced to one in which the new matrix A is of full rank, and the value of m reduced. So, if the theorem is proved for the case where A is of rank $m \leqslant n$, all cases are covered.

To prove the result, suppose that the non-zero elements of a particular solution x are just $x_1, x_2, ..., x_r$, and that $r > m$. (In this context, the convention is to use the term "solution" for any x feasible for the problem, and the term "optimal solution" for the particular feasible x which maximize f.) Now, by solving the system (4.32) for $x_1, x_2, ..., x_m$ we can rewrite the constraints (4.32) in the form

$$x_j = d_j - \sum_{k=m+1}^{n} \alpha_{jk} x_k \quad (j = 1, 2, ..., m). \tag{4.35}$$

The two formulations (4.32) and (4.35) of the constraint are equivalent, and all x satisfying (4.35) also satisfy (4.32).

The solution will now be improved if x can be varied, consistently with (4.35), in such a way as to increase f. This can be done most easily by regarding $x_{m+1}, x_{m+2}, ..., x_n$ as independent variables, which can be chosen freely (in $x \geqslant 0$), the values of $x_1, x_2, ..., x_m$ then being determined by (4.35). Substitution of expressions (4.35) for $x_1, ..., x_m$ into $c^T x$ yields an analogous expression for $f(x)$ in forms of these independent variables:

$$f(x) = \sum_{1}^{m} c_k d_k - \sum_{m+1}^{n} \gamma_k x_k. \tag{4.36}$$

In particular, consider a variation of x_r: this will affect $x_1, \ldots, x_m$, but will leave all the zero elements $x_{r+1}, \ldots, x_n$ unaffected. If x_r is modified to $x_r + \theta$, where θ has the opposite sign to γ_r (either sign if $\gamma_r = 0$), then $f(x)$ will certainly not be decreased by the change. If θ is increased sufficiently in magnitude, then ultimately either x_r or one of $x_1, x_2, \ldots, x_m$ will become zero. At that point no further increase in θ is possible, or an element of x will become negative, and the solution infeasible. However, the objective has been attained: the number of non-zero elements in x has been decreased by one, with no decrease (and, in most cases, a positive increase) in f. The process can be repeated as long as $r > m$. The theorem is thus proved.

Indeed, it is seen that this same technique can be used to improve a solution which is already basic. Consider an initial basic solution, $\bar{x}$. The relation corresponding to (4.35) must then be

$$x_j = \bar{x}_j - \sum_N \alpha_{jk} x_k \quad (j \in B). \tag{4.37}$$

Here B is the *basis*, the set of j for which $\bar{x}_j > 0$. The summation over k in N covers the *non-basic* variables: those k for which $\bar{x}_k = 0$. The absolute term d_j of (4.35) is now identifiable with $\bar{x}_j$, because we know that x_j equals $\bar{x}_j$ when the non-basic variables are set equal to zero, by definition.

Suppose that (4.36) is correspondingly written

$$\begin{aligned} f(x) &= \sum c_k \bar{x}_k - \sum_N \gamma_k x_k \\ &= f(\bar{x}) - \sum_N \gamma_k x_k. \end{aligned} \tag{4.38}$$

Then, as long as $\gamma_k < 0$ for some k, say k_0, the solution can be improved. For x_{k_0} can be modified to a positive value, say θ, and θ can be increased, simultaneously increasing f, until one of the basic variables becomes zero, say x_{j_0}. Since under the variation $x_{k_0} \to \theta$ the basic variables change by the rule

$$x_j \to \bar{x}_j - \alpha_{jk_0} \theta. \tag{4.39}$$

it is seen that j_0 must be a value of j minimizing $\bar{x}_j / \alpha_{jk_0}$ in B.

Now, the effect of this variation is that x_{k_0} has become positive, and x_{j_0} zero. That is, there has been a change to a new basis, in which k_0 replaces j_0, and the solution has thereby been improved. This improvement technique is just the *simplex method*, and it leads one to an absolute maximum of f.

Theorem (4.3)

Let the basis-changing technique just described be continued until all γ_k in representation (4.38) are non-negative. This stage is reached in a finite number of steps, and, when it is reached, a solution is obtained which maximizes $f(x)$ absolutely in the feasible region.

There are a finite number of basic solutions—indeed, not more than $\binom{n}{m}$— and the sequence of basic solutions generated by this technique can never cycle back on itself (since f increases at each stage). The sequence must thus terminate in at most $\binom{n}{m}$ steps. The condition $\gamma_k \geqslant 0$ $(k \in N)$ is necessary at termination, otherwise further improvement would be possible. It is also sufficient, for it implies that, for any feasible x,

$$f(x) = f(\bar{x}) - \sum \gamma_k x_k \leqslant f(\bar{x}), \qquad (4.40)$$

so that $\bar{x}$ provides a global optimum.

The argument of Theorem (4.3) is lacking only in one respect. Suppose that, at some change in the basis, more than one element in the old basis becomes zero. The new basis then has fewer than m elements, and is *degenerate*. In order to continue the basis-changing algorithm of the theorem, one must expand the basis to a full set of m elements, which then necessarily includes the labels k of some zero x_k. If these additional elements are chosen inappropriately, there may be no perturbation (4.39) which leads to an actual increase in f; time can thus be spent cycling ineffectually between such bases, looking for one from which a real improvement is possible. There are guides for a correct choice of basis (see Karlin (1959), Vol. I, p. 167). These all depend upon the fact that the degeneracy can be removed by a perturbation of the problem, in which some element of A, b or c is modified by an arbitrarily small amount.

Exercise

(4.10) Find linear problems other than the allocation problem which are reducible to the standard form of this section. Determine the method of standardization for the problems of Section (4.4).

4.6 THE SIMPLEX METHOD: THE TABLEAU

The calculations of the simplex method can be schematized by use of the *tableau* and its transformations. Suppose, for concreteness, that one is considering a basic solution $\bar{x}$ in which $x_1, x_2, \ldots, x_m$ are the basic variables;

$x_{m+1}, \ldots, x_n$ the non-basic variables. Then the tableau is defined as the array:

		$m+1$	$\ldots$	n
1	$\bar{x}_1$	$\alpha_{1,m+1}$	$\ldots$	α_{1n}
2	$\bar{x}_2$	$\alpha_{2,m+1}$	$\ldots$	α_{2n}
$\vdots$	$\vdots$	$\vdots$	$\vdots$	$\vdots$
m	$\bar{x}_m$	$\alpha_{m,m+1}$	$\ldots$	α_{mn}
	$f(\bar{x})$	γ_{m+1}	$\ldots$	γ_n

That is, along the left are listed the *labels* of the basic variables, either in the form of the appropriate subscripts k, or of the full symbol x_k, or of a verbal description of the variables. Along the top are listed the labels of the non-basic variables. The body of the tableau contains the numerical values of the basic variables, of the function f, and of the coefficients α_{jk} and γ_k as indicated.

Now, in changing to an improved basis, a value k is first selected for which γ_k is negative; k_0, say. This is usually a value for which γ_k is most negative; a reasonable, although not necessarily optimal, procedure. Having chosen a column, label k_0, one selects a row label j_0 from among the values of j minimizing $\bar{x}_j/\alpha_{jk_0}$. The point j_0, k_0 thus determined in the tableau is termed the *pivot*.

According to the calculations of the previous section, a new basis is now adopted in which k_0 replaces j_0. All the entries in the tableau will thereby be changed, but the transformation is a simple one.

Theorem (4.4)

Suppose j_0, k_0 has been chosen as pivot. Then the tableau transforms according to the following rules:

(i) *The labels j_0 and k_0 are interchanged.*

(ii) *The elements α_{jk} undergo the transformations*

$$\alpha_{j_0k_0} \to 1/\alpha_{j_0k_0}, \tag{4.41}$$

$$\alpha_{j_0k} \to \alpha_{j_0k}/\alpha_{j_0k_0} \quad (k \neq k_0), \tag{4.42}$$

$$\alpha_{jk_0} \to -\alpha_{jk_0}/\alpha_{j_0k_0} \quad (j \neq j_0), \tag{4.43}$$

$$\alpha_{jk} \to \alpha_{jk} - \alpha_{j_0k}\,\alpha_{jk_0}/\alpha_{j_0k_0} \quad (j \neq j_0, k \neq k_0). \tag{4.44}$$

(iii) *The entries $\bar{x}_j, f(\bar{x}), \gamma_k$ transform just as if they constituted another column and row of the α matrix.*

For, since x_{j_0} and x_{k_0} are now changing places,

$$x_{j_0} = \bar{x}_{j_0} - \sum \alpha_{j_0 k} x_k \tag{4.45}$$

must be inverted to

$$x_{k_0} = \frac{1}{\alpha_{j_0 k_0}} \left[\bar{x}_{j_0} - x_{j_0} - \sum_{k \neq k_0} \alpha_{j_0 k} x_k \right], \tag{4.46}$$

and (4.46) replaces (4.45) in the set of relations (4.37). Examination of the coefficients of x_{j_0} and x_k $(k \neq k_0)$ in (4.46) confirms rules (4.41) and (4.42).

Inserting expression (4.46) into the general relation (4.37) for $j \neq j_0$, we obtain the representation of x_j in terms of the new non-basic variables:

$$x_j = \bar{x}_j - \sum_{k \neq k_0} \alpha_{jk} x_k - (\alpha_{jk_0}/\alpha_{j_0 k_0}) \left[\bar{x}_{j_0} - x_{j_0} - \sum_{k \neq k_0} \alpha_{j_0 k} x_k \right]. \tag{4.47}$$

Examination of the coefficients of x_k in this expression confirms transformations (4.43) and (4.44). Identification of the constant terms in (4.46) and (4.47) similarly reveals that the numerical values of the new basic variables are

$$\bar{x}_{k_0} \to \bar{x}_{j_0}/\alpha_{j_0 k_0},$$

$$\bar{x}_j \to \bar{x}_j - \alpha_{jk_0} \bar{x}_{j_0}/\alpha_{j_0 k_0} \quad (j \neq j_0).$$

That is, the $\bar{x}$ column of the tableau has indeed transformed according to rules (4.42) and (4.44), just as though it were another column of the α-matrix. Correspondingly, substitution of (4.46) in (4.38) yields

$$f(\bar{x}) \to f(\bar{x}) - \gamma_{k_0} \bar{x}_{j_0}/\alpha_{j_0 k_0},$$

$$\gamma_{k_0} \to -\gamma_{k_0}/\alpha_{j_0 k_0},$$

$$\gamma_k \to \gamma_k - \alpha_{j_0 k} \gamma_{k_0}/\alpha_{j_0 k_0} \quad (k \neq k_0),$$

so that rule (iii) holds also for the $f(\bar{x}), \gamma$ row of the tableau.

For a numerical example, consider the case of a company which can manufacture three fuel oils from three fuel stocks, the proportional compositions of the fuels being

	Stock 1	2	3
Fuel 1	0.3	0.5	0.2
Fuel 2	0.4	0.3	0.3
Fuel 3	0.5	0.4	0.1

The stocks are available in amounts of 15, 20 and 12 tons respectively, and the profit on each ton of fuel is 3.0, 2.0 and 2.5 respectively.

If x_k is the amount of fuel k manufactured, then one has the problem of choosing the x_k so as to maximize

$$f = 3x_1 + 2x_2 + 2.5x_3,$$

subject to

$$0.3x_1 + 0.4x_2 + 0.5x_3 \leqslant 15,$$
$$0.5x_1 + 0.3x_2 + 0.4x_3 \leqslant 20,$$
$$0.2x_1 + 0.3x_2 + 0.1x_3 \leqslant 12.$$

Multiplying these relations by 10 to clear of decimals and adding slack variables, one obtains the equality constraints

$$3x_1 + 4x_2 + 5x_3 + x_4 = 150,$$
$$5x_1 + 3x_2 + 4x_3 + x_5 = 200,$$
$$2x_1 + 3x_2 + x_3 + x_6 = 120.$$

The simplest course is to take $(0, 0, 0, 150, 200, 120)$ as the first basic solution, yielding the following tableau:

		x_1	x_2	x_3
x_4	150	3	4	5
x_5	200	5	3	4
x_6	120	2	3	1
	0	-3	-2	-2.5

Since the numerically largest negative γ entry occurs in the x_1 column, this is taken as the pivot column. The x_5 row must then be the pivot row, since $200/5$ is the smallest of $150/3$, $200/5$ and $120/2$. The transformation rules then lead to the tableau

		x_5	x_2	x_3
x_4	30	$-3/5$	$11/5$	$13/5$
x_1	40	$1/5$	$3/5$	$4/5$
x_6	40	$-2/5$	$9/5$	$-3/5$
	120	$3/5$	$-1/5$	$-1/10$

Taking the x_4/x_2 element as the next pivot, one derives the transformed tableau

		x_5	x_4	x_3
x_2	$13\frac{7}{11}$	$-3/11$	$5/11$	$13/11$
x_1	$31\frac{9}{11}$	$4/11$	$-3/11$	$1/11$
x_6	$15\frac{5}{11}$	$1/11$	$-9/11$	$-30/11$
	$122\frac{8}{11}$	$6/11$	$1/11$	$3/22$

Since all γ_k are now positive, the solution has been reached and iteration ends. The optimal policy is to blend $13\frac{7}{11}$ tons of fuel 1, $31\frac{9}{11}$ tons of fuel 2 and none of fuel 3, leaving a surplus of $1\frac{6}{11}$ tons of stock 3.

4.7 RELATIONSHIP OF THE SIMPLEX METHOD TO THE DUAL PROBLEM

In Section (4.5) the simplex method was developed from first principles. However, it has a natural relationship to the Lagrangian theory established earlier, and this relation can be helpful in special cases.

Suppose that $\bar{x}$ is a basic solution (i.e. a basic feasible vector) for the standard problem posed in Section (4.5). Let the column vector x be partitioned into $\begin{pmatrix} x_B \\ x_N \end{pmatrix}$ where x_B and x_N are the vectors of basic and non-basic variables respectively. Let the corresponding partitions of A and c be denoted $(A_B\,A_N)$ and $\begin{pmatrix} c_B \\ c_N \end{pmatrix}$.

Theorem (4.5)
  Define the vector

$$y = \partial f(\bar{x})/\partial b, \tag{4.48}$$

interpretable as the effective price vector if the basis B is held fixed. Then

$$y^{\mathrm{T}} = c_B^{\mathrm{T}} A_B^{-1} \tag{4.49}$$

and

$$y^{\mathrm{T}} A - c^{\mathrm{T}} = (0, \gamma^{\mathrm{T}}), \tag{4.50}$$

where γ is the vector of coefficients in representation (4.38). Thus, y is feasible for the dual problem if and only if $\bar{x}$ solves the primal (when $\gamma \geqslant 0$). In this case, y solves the dual.

Recall the interpretation of $y^T A - c^T$ as the row vector of "net unit costs" for the n activities: costs of resources used (on prices y) less return received, per unit intensity. It is seen from relation (4.50) that, if one adopts a particular basis, then these net costs are zero for activities in the basis. That is, the prices effective for this basic solution can be determined by requiring that those activities which are undertaken just break even. The net unit costs for non-basic activities are then just the γ_k of representation (4.38), and this is where one perceives the relation between the simplex method, and the use of the effective price vector y to deduce a costing of the various activities.

It is known from Section (4.5) that, if $\gamma_k < 0$, then activity k can advantageously be brought into the basis, and that, conversely, no further improvement is possible if $\gamma \geqslant 0$. One now sees that the inequality $\gamma_k < 0$ is equivalent to the statement that $\sum_j y_j a_{jk} < c_k$, i.e. that, on current prices y, activity k has negative unit cost and is potentially profitable. The changes of basis which appear advantageous in the simplex method, and which ultimately bring the allocation x to optimality for the primal problem, are then just those which appear advantageous when activities are costed on current effective prices y. Furthermore, when optimality has been reached, and $\gamma \geqslant 0$, then $y^T A \geqslant c^T$, so that y is feasible for the primal. The changes of basis of the simplex method which move x towards optimality thus move y towards feasibility: the two states are reached simultaneously, and y then in fact solves the dual.

The effective prices y did not appear overtly in the simplex method; one worked directly with the combinations $\gamma_k = \sum_j y_j a_{jk} - c_k$. However, for some problems (such as the transportation problem of the next section) it is advantageous to calculate the dual variables y explicitly at each stage, and to use them to cost activities explicitly.

To prove the theorem, rewrite (4.32) as

$$A_B x_B + A_N x_N = b. \tag{4.51}$$

If the actual numerical values in the basic solution $\bar{x}$ are considered, when $x_B = \bar{x}_B$ and $x_N = 0$, then it follows from (4.51) that

$$\bar{x}_B = A_B^{-1} b \tag{4.52}$$

Thus

$$f(\bar{x}) = c_B^T \bar{x}_B = c_B^T A_B^{-1} b, \tag{4.53}$$

so that the vector (4.48) is indeed given by (4.49).

Solving (4.51) for x_B, one obtains the representation (4.37) in the form

$$x_B = \bar{x}_B - A_B^{-1} A_N x_N. \tag{4.54}$$

Substituting this into $f(x)$, one deduces representation (4.38) in the form

$$
\begin{aligned}
f(x) &= c_B{}^{\mathrm{T}} x_B + c_N{}^{\mathrm{T}} x_N \\
&= c_B{}^{\mathrm{T}} (\bar{x}_B - A_B^{-1} A_N x_N) + c_N{}^{\mathrm{T}} x_N \\
&= f(\bar{x}) - (y^{\mathrm{T}} A_N - c_N{}^{\mathrm{T}}) x_N.
\end{aligned} \tag{4.55}
$$

Identification of the vector coefficient of x_N in (4.38) and (4.55) shows that $\gamma^{\mathrm{T}} = y^{\mathrm{T}} A_N - c_N{}^{\mathrm{T}}$; from this and equation (4.49) assertion (4.50) follows.

The vector y will be feasible for the dual if and only if $y^{\mathrm{T}} A - c^{\mathrm{T}} \geqslant 0$ (since (4.32) is an equality, there is no other constraint on the value of y). So, the second-last assertion of the theorem follows.

Note now that

$$
c^{\mathrm{T}} \bar{x} = c_B{}^{\mathrm{T}} A_B^{-1} b = y^{\mathrm{T}} b. \tag{4.56}
$$

However, if $\bar{x}$ and y are feasible for primal and dual respectively, then

$$
c^{\mathrm{T}} \bar{x} \leqslant U(b) \leqslant y^{\mathrm{T}} b. \tag{4.57}
$$

The equality in (4.56) thus implies that $\bar{x}, y$ solve primal and dual respectively if they are feasible for their respective problems; the final assertion of the theorem thus follows.

There would appear to be no difficulty about the existence of the derivative (4.48), since $f(\bar{x})$ has the explicit linear form (4.53) as a function of b. This would certainly be true if $\bar{x}_B \geqslant 0$, when solution (4.52), deduced on the assumption of a fixed basis B, remains feasible (i.e. none of the $\bar{x}_k$ become negative) as b is varied in a neighbourhood of the given value. However, if some elements of $\bar{x}_B$ are zero (i.e. the basis is degenerate) then some infinitesimal perturbations in b will necessitate a change of basis if the solution is to remain feasible. In this case, solution (4.52) can hold only for certain b in the neighbourhood of the prescribed value, and the vector y of (4.49) has a price interpretation only for certain directions of b-perturbation. However, the theorem and its proof remain valid.

4.8   THE TRANSPORTATION PROBLEM

The transportation problem of Section (4.4b) has a special structure, but rather many variables. This makes a tableau presentation of the computational solution clumsy, although the simplex method is itself still appropriate. In fact, the problem is ideally suited to the formulation of the last section, in which the simplex method is seen as a casting back and forth between the allocation variables of the primal and the price variables of the dual.

Suppose that total demand exactly equals total supply,

$$
\sum r_j = \sum s_k, \tag{4.58}
$$

so that relations (4.21) and (4.22) necessarily become equalities. If demand exceeds supply the problem has no solution; if supply exceeds demand the equality (4.58) can always formally be achieved by the introduction of a dummy destination, whose requirement equals the surplus, and to which shipping costs are zero. This dummy destination provides the slack variables appropriate to the problem.

If Lagrangian multipliers α_j and β_k are introduced for constraints (4.21) and (4.22) respectively, then these must have the interpretations

$$\alpha_j = -\frac{\partial U(r,s)}{\partial r_j}, \tag{4.59}$$

$$\beta_k = \frac{\partial U(r,s)}{\partial s_k}; \tag{4.60}$$

literally, if the derivatives exist; in the generalized sense of Theorem (3.9) if they do not. That is, α_j would be a fair marginal purchase price (per unit amount) for the commodity at depot j, and β_k a fair marginal selling price at destination k. Care must be taken in talking about "prices", in fact these prices refer only to the transportation cost component of the commodity, and not to its intrinsic value at sources or market value at the destinations. More specifically, if the supply–demand equality (4.58) is imposed, it will be seen that the α_j and β_k are indeterminate up to an additive constant, so that only price differences are significant. Thus, if demand at destination k and supply at source j were both increased by δ, then total minimal shipping costs would increase by $(\beta_k - \alpha_j)\,\delta + o(\delta)$. Similarly, if a demand δ were transferred from destination k to destination k', costs would increase by $(\beta_{k'} - \beta_k)\,\delta + o(\delta)$. This is the significance of the multipliers, to be kept in mind when they are subsequently referred to as effective purchase or selling prices.

The problem dual to the original (in the case of equality constraints) is the maximization of

$$\sum \beta_k s_k - \sum \alpha_j r_j \tag{4.61}$$

subject to

$$\beta_k - \alpha_j \leqslant c_{jk}. \tag{4.62}$$

That is: prices to be such that the value of sales less the value of purchases is to be maximal, subject to the requirement that no individual route shows a profit.

The algorithm is then as follows. Since, in the case of equality, conditions (4.21) and (4.22) amount to $m+n-1$ distinct constraints, a basic solution will have no more than $m+n-1$ non-zero x_{jk}, and exactly $m+n-1$ if the problem is non-degenerate.

Having found such a solution, determine the effective prices α_j, β_k from the principle that basic activities should just break even. That is,

$$\beta_k - \alpha_j = c_{jk} \tag{4.63}$$

for j, k in the basis; for those routes jk which receive a positive loading. Relation (4.63) only supplies $m + n - 1$ equations for these $m + n$ prices, but an arbitrary value can be assigned to one of them (say, $\alpha_1 = 0$: see Exercise (4.12)).

As in Section (4.7), the effective prices α, β will not be feasible values for the dual problem unless the suggested transport pattern is in fact optimal. Thus, if the pattern is not optimal, inequality (4.62) will be violated for some j, k (necessarily corresponding to a route not included in the trial basic solution) so that traffic on such a route would in fact be profitable, on current prices. One then increases traffic on such a route, decreasing the existing route loadings correspondingly, until some loading in the old basis becomes zero. A change of basis for the better has thus been achieved, and the process can be repeated.

The calculation can be schematized very concisely; the schematization is illustrated for a case with three sources and three destinations.

Sources

	5	6	3	3
Destinations	3	10	4	9
	1	3	7	14
	10	8	8	

The unit transportation costs c_{jk} are set out in a two-way array, the convention being to allocate sources to columns and destinations to rows. This is bordered with a lowest row of supplies, r_j, and a rightmost column of requirements, s_k.

The next step is to determine an initial basic solution. One way to do this is the "northwest" method. As much as possible of the supply from source 1 is allocated to destination 1, then to destination 2, and so on until exhausted. One then continues by allocating the supply from source 2 in the same way. In this way one progresses through all sources and destinations, and ends without surplus or deficit if relation (4.58) is fulfilled. In the following table we have entered the basic solution thus obtained, and have also entered the transport cost c_{jk} in the south-east corner of each cell.

	0	−7	−11	
5	3−θ 5	 6	θ · 16 3	
3	7+θ 3	2−θ 10	 4	Cost = 130.
−4	 1	6+θ 3	8−θ 7	

Effective prices α_j, β_k are now determined by setting $\alpha_1 = 0$ and then using relation (4.63) for all cells j, k in the basis. These prices are entered to the top and left of the table, as indicated.

One then looks for "profitable routes", which violate relation (4.62). The most flagrant violation is in the third cell of the first row, for which $c_{jk} = 3$ but $\beta_k - \alpha_j = 16$. (The test calculations of $\beta_k - \alpha_j$ can conveniently be entered in the north-east corner of each cell.) The loading is then increased on this route by θ, and other loadings are adjusted accordingly to preserve row and column totals. Note that these adjustments must be so made that only *one* loading not in the old basis is increased, since it is desired to keep the solution basic. From the θ entries in the last table one sees that θ can be increased to $\theta = 2$, when the (2, 2) loading vanishes, and one deduces the new table.

	0	6	2	
5	1−θ 5	 6	2+θ 3	
3	9 3	 10	 4	Cost = 104.
9	θ · 9 1	8 3	6−θ 7	

Continuing as indicated, one obtains the two following transformations of the table.

	0	−2	−6
−3	5	6	3 ⌐ 3
3	9−θ ⌐ 3	10	θ ⌐ 9 ⌐ 4
1	1+θ ⌐ 1	8 ⌐ 3	5−θ ⌐ 7

Cost = 96.

	0	−2	−1
2	2 ⌐ 5	4 ⌐ 6	3 ⌐ 3 ⌐ 3
3	4 ⌐ 3 ⌐ 3	5 ⌐ 10	5 ⌐ 4 ⌐ 4
1	6 ⌐ 1 ⌐ 1	8 ⌐ 3 ⌐ 3	2 ⌐ 7

Cost = 71.

For the last table, inequality (4.62) is satisfied for all cells, and the loading pattern indicated by the north-west cell entries is consequently optimal.

An interesting variant of the problem would be the situation in which the company wishing to transport the commodity owned its own means of transport (ships, etc.) so that allocation of shipping would be determined by best use of its own limited transport resources, rather than by minimization of the costs (on prices c_{jk}) of hiring transport. However, the Lagrangian multipliers associated with shipping constraints would then generate "shadow" transport prices, c_{jk}.

The costs c_{jk} for the case of a company which hires transport could be regarded as the shadow prices for a larger problem, embracing the whole transport system. However, if it is desired to consider, not a whole system, but merely the operations of a single company, then the prices provide a means of coupling this company's operations to the external world, and of enabling one to study the sub-problem (operations of the single company) in isolation.

Exercises

(4.11) Suppose that a problem is subject to a set of m mutually consistent linear equality constraints $Ax = b$, of which only $r < m$ are linearly independent. Show that the Lagrangian multipliers y_j associated with the redundant constraints are indeterminate in value. More specifically, if a vector ξ exists such that $\xi^T A = 0$, $\xi^T b = 0$ then y may arbitrarily be modified to $y + \lambda \xi$, for any scalar λ.

(4.12) Show that, if (4.58) holds, then (4.21) and (4.22) amount only to $m + n - 1$ functionally independent constraints, and that α_j, β_k are then determined only to within an additive constant. Interpret. Show that, in the case $\sum r_j > \sum s_k$, at least one α_j will be zero.

(4.13) The multipliers α, β associated with the inequality constraints (4.21) and (4.22) must be non-negative, as is evident from the differential interpretations (4.59) and (4.60). On the other hand, these constraints are effectively equality constraints if the balance relation (4.58) holds, and α, β are then unrestricted. Reconcile these two assertions. If the constraint (4.58) is imposed then the r_j and s_k can no longer be as varied independently, and the simple differential interpretations (4.59) and (4.60) can no longer be valid. What can be said? (assuming U differentiable).

CHAPTER 5

Some Particular Linear Problems

Some linear problems are sufficiently individual in structure that one can continue their analytic treatment past the point reached in Chapter 4. This chapter will be devoted to examples of this type, interesting for one reason or another. Some of these have the full LP form, others do not, in that the basic domain $\mathscr{X}$ is non-conical.

The first three sections of the chapter are independent of one another, and of the remaining sections.

5.1 LINEAR MODELS OF ECONOMIC DEVELOPMENT

A very interesting generalization of the allocation problem of Section (4.3) is a multi-stage allocation process, in which part at least of the goods produced in any stage of the process is reinvested to constitute part of the raw materials of the next stage, and so on. This provides a model for a dynamic economy or technology, in which one can examine the allocation of products at every stage, between consumption and different modes of reinvestment, which is optimal in that it maximizes some measure of overall utility.

Such dynamic optimization processes lie somewhat outside the limits chosen for this book, but this particular example is such an immediate and interesting extension of the single-stage problem already considered that the excursion is worth while.

No explicit distinction will be drawn between goods which occur naturally or are manufactured; any such distinction will be implicit in the availabilities, production coefficients, etc. assigned. Nor will any distinction be made in the first instance between activities which are productive and those which merely give satisfaction, without further product. These rather inclusive view-points make for a more compact model. They may also correspond to reality: many of the apparently non-utilitarian activities of a leisured and cultured society are in the long run productive, in that they provide the path of development

97

to a higher form of society. Conversely, a utilitarian activity can give satisfaction, even if none of its product is siphoned off for "consumption".

As before, let a_{jk} denote the amount of commodity required for activity k at unit intensity, b_j the amount of commodity j naturally available per stage, and c_k the utility derived per stage from unit intensity of activity k. Let b_{jk} (the element of a matrix B) denote the amount of commodity j *produced* by unit intensity of activity k. If $x(t)$ is the intensity vector at stage t, where t runs through the integers, then we have the material balance inequality

$$Ax(t) \leqslant b + Bx(t-1). \tag{5.1}$$

It will be supposed that the aim is to choose the intensity sequence $\{x(t)\}$ subject to (5.1), to the condition

$$x(t) \geqslant 0, \tag{5.2}$$

and to prescription of $x(0)$ so as to maximize the total utility

$$f(x) = \sum_{t=1}^{N} c^{\mathrm{T}} x(t). \tag{5.3}$$

As indicated above, consumption does not appear explicitly, because this is regarded as just another activity, which may or may not be productive, according to whether elements of B for the appropriate k and some j are positive or zero.

The quantities A, B, b and c can all be made t-dependent (time-varying) but we shall content ourselves with a system of constant structure.

The programming problem defined by (5.1)–(5.3) is a linear one, although of a special recursive form. A vector $y(t)$ of non-negative Lagrangian multipliers can be associated with the vector inequality (5.1): this will have the interpretation of a shadow price vector for goods just before the tth stage of manufacture. The dual problem is: minimize $\sum_1^N y(t)^{\mathrm{T}} b$ subject to

$$y(t)^{\mathrm{T}} A - y(t+1)^{\mathrm{T}} B \geqslant c^{\mathrm{T}} \quad (t = 1, 2, ..., N), \tag{5.4}$$

$$y(N+1) \quad = 0. \tag{5.5}$$

That is, total resources supplied over the period $1 \leqslant t \leqslant N$ shall have minimal value, subject to the requirement that, for each activity at each stage, the difference between the values of goods consumed and produced shall not be less than the utility derived from that activity. In other words, no activity ever shows a "profit". Equality will hold in every row of (5.4) for which the corresponding element of $x(t)$ is strictly positive. Furthermore, by (5.5), all goods shall have zero value at the end of the programme, reflecting the fact that criterion (5.3) only assigns value to activities up to time N.

The interesting question is now: does an asymptotic pattern of allocation emerge as N becomes large? A thorough treatment of this question would require more space than is appropriate here, but the principal ideas can be sketched.

Define

$$\rho = \max_{x>0} \min_{y>0} \left(\frac{y^T Bx}{y^T Ax}\right). \tag{5.6}$$

Then ρ is the greatest rate of economic expansion that can be achieved in the absence of external resources, i.e. if $b = 0$. Here, "external resources" is understood not to include those resources which may be taken as freely available in unlimited quantities (e.g. air, space, water), if there are any such, because these will not be included in the material balance (5.1).

The definition of ρ by a simultaneous maximization and minimization (cf. (4.18)) leads again to the formulation of a game, treated in Section (10.4). The ideas are no more than sketched in this section, but the argument behind some of the assertions made is given in Exercises (5.2) and (5.3). See also Exercise (10.22).

Suppose now that $\rho \geqslant 1$. Then, subject to non-degeneracy requirements and a suitable value of $x(0)$, the economy can grow at an exponential rate (or just sustain itself, in the case $\rho = 1$) without use of external resources other than those which are freely available. That is, the economy can itself generate all critical commodities needed in sufficient quantities to ensure economic survival and growth. If $\bar{x}$ and $\bar{y}$ are the values of x and y giving the extremum in (5.1). One finds that, for values of t remote both from 0 and N,

$$\left.\begin{array}{l} x(t) \sim \text{const } \rho^t \bar{x}, \\ y(t) \sim \text{const } \rho^{-t} \bar{y}. \end{array}\right\} \tag{5.7}$$

That is, allocation has a constant structure in that $x(t)$ and $y(t)$ are proportional to fixed vectors multiplied by an exponential time factor. Intensities rise at an exponential rate (reflecting the overall exponential growth) and shadow prices decrease correspondingly (because an injection of goods would be more valuable early in the programme).

However, the notion of economic growth in a vacuum seems fallacious. There is a constant dependence on the environment, and sooner or later some external resource will prove critical, in that its limited rate of supply will limit the scale of the economy. Furthermore, the "expansionist" economy leads to strange allocations, in which the emphasis is all on overtly productive activities (i.e. maximal reinvestment, minimal consumption). This is because an asset reinvested today, rather than consumed, will grow exponentially with time to produce a much greater asset (and potential utility) later. Indeed, the

whole question of reaping the reward of all this activity is postponed indefinitely, "to the end of the plan". This is demonstrated by the fact that (apart
from non-degeneracy requirements) the optimal expansionist programme is
independent of c, implying a preoccupation with growth for its own sake,
rather than with the relative evaluation of different activities.

Suppose now that $\rho < 1$. Then economic survival is impossible, if external
sources of supply are cut off. This will always be the case, if dependence on the
the environment and limitations of the environment are frankly recognized.
In this case, with a given b (i.e. a given environment), the economy will
expand up to a point where this environment is fully and efficiently utilized,
and an equilibrium situation will be reached. Despite the fact that the
economy is not a self-sustaining one in isolation, considerable "amplification"
of environmental resources is possible (by a factor of the order of $(1-\rho)^{-1}$).
If x and y are the equilibrium values of $x(t)$ and $y(t)$, then x will maximize
$c^T x$ subject to $x \geqslant 0$ and

$$(A-B)x \leqslant b. \tag{5.8}$$

The programme now depends explicitly on b and c; i.e. takes account of
environmental limitations, and of the relative values of different activities.
The equilibrium shadow price vector y will minimize $y^T b$ subject to $y \geqslant 0$ and

$$y^T(A-B) \geqslant c^T. \tag{5.9}$$

A defect of this model is the assumption of linear utilities: more particularly
of additive utilities. In (5.3) utilities are added over different time-periods,
and over different activities within a time-period. But, for example, it is
almost certainly true that a number of different activities must be carried on
at certain minimal levels (i.e. different goods must be supplied for consumption at certain minimal rates) if life is to be possible, and this fact should be
represented in the utility function. In other words, one cannot always
substitute one good for another (e.g. apricots for coal), as the adoption of a
linear utility function implicitly assumes. Section (6.1) will go some distance
towards meeting this point, by considering nonlinear utility functions.

Exercises

(5.1) Consider the case of one commodity and two activities, one
productive and one unproductive, so that condition (5.1) reduces to the
scalar inequality,

$$a_{11} x_1(t) + a_{12} x_2(t) \leqslant b_1 + b_{11} x_1(t-1),$$

all coefficients (including c_1 and c_2) being supposed positive. Determine ρ
and calculate the optimal steady-state allocation. Solve the multi-stage

allocation problem for finite N, and relate the results to the steady-stage solution.

(5.2) Characterize ρ as the largest value for which $(B - \rho A)x \geqslant 0$ for some $x > 0$.

(5.3) Assuming positivity of $y^T Ax$, show that $\rho < 1$ iff $(B - A)x < 0$ for all $x > 0$, and that $\rho \geqslant 1$ iff $(B - A)x \geqslant 0$ for some $x > 0$. Suppose that, in this latter case, $c^T x > 0$ for the x vector found. Show then that an attempt to solve the equilibrium optimal problem associated with (5.8) leads to $U = +\infty$.

(5.4) The term "degeneracy" was used in the text to cover a number of rather special circumstances. For example, it may be that certain goods can be produced in ever-increasing amounts in an isolated economy, but that these goods carry no utility—"economic expansion" in such a case is an empty term. Alternatively, goods may fall into two clasees, I and II, such that each activity involves inputs and outputs only of goods of one class. Isolated expansion may be possible for an economy based on goods of class I, but there may be goods only of class II to begin with, so the path to expansion is barred.

Formalize these situations in terms of properties of the matrices A, B, b, c.

5.2 ALLOCATION WITH BOUNDED INTENSITIES; HYPOTHESIS-TESTING AND THE NEYMAN–PEARSON LEMMA

Consider again the linear allocation problem of Section (4.3), but now take the basic domain $\mathcal{X}$ as being

$$0 \leqslant x \leqslant M, \qquad (5.10)$$

rather than the whole positive orthant. That is, the intensity of activity k is not allowed to exceed a level M_k $(k = 1, 2, \ldots, n)$. This ceiling may reflect physical limitations, a point of decreasing return, or a desire to avoid the risks of over-concentrated investment.

Since $\mathcal{X}$ is no longer conical the complete LP formulation of Section (4.1) does not hold. However, it is convex and polyhedral, so that $U(\xi)$ is concave and piecewise linear. If $U(\xi)$ is in addition finite for all attainable ξ in some neighbourhood of b (certainly true if M is finite) then the strong Lagrangian principle is certainly valid, with a finite multiplier: see (iiib) of Theorem (3.9). Taking account then of the restrictions $Ax \leqslant b$ by a multiplier y, one sees that the optimal intensities satisfy

$$\bar{x}_k = \begin{Bmatrix} 0 \\ \text{indeterminate} \\ M_k \end{Bmatrix} \quad \text{according as} \quad (y^T A)_k - c_k \begin{Bmatrix} > \\ = \\ < \end{Bmatrix} 0, \qquad (5.11)$$

and that y is determined from the dual problem: minimize

$$F(y) + y^T b = \sum_k M_k [c_k - (y^T A)_k]_+ + y^T b \tag{5.12}$$

in $y \geqslant 0$. If all elements M_k are bounded, then so is $\bar{x}$, and the dual problem certainly has a solution.

The non-conical nature of $\mathscr{X}$ reflects itself in the nonlinear character of the dual. If the M_k are allowed to tend to infinity we recover the pattern of Section (4.3).

The simplex method can still be used, in adapted form, to provide a computational algorithm which resolves the indeterminacy in (5.12). The variable not in the basis must be divided into two groups, those taking the "floor" value $x_k = 0$, and those taking the "ceiling" value $x_k = M_k$. The variables of intermediate value constitute the basis. Suppose the problem normalized by the introduction of slack variables, so that constraints take the form of m linearly independent equality constraints. Then, as in Theorem (4.2), the basis need contain not more than m variables. The basic variables and criterion function are then represented in terms of the non-basic variables, as in (4.37) and (4.38), and the solution improved by perturbation of one of the non-basic variables (to be lifted from zero or reduced from M_k, as appropriate) until the basis changes. The calculation terminates when all the γ_k in (4.38) are non-negative for "floor" variables, non-positive for "ceiling" variables.

An interesting one-constraint version of this situation occurs in the Neyman–Pearson theory of hypothesis-testing. Suppose that, on the evidence of an experimental observation ω, one wishes to test the validity of a hypothesis H_0, the *null hypothesis*. One can do this virtually only by asking whether H_0 accounts better for the experimental finding than other conceivable hypotheses. Suppose that a particular alternative is in mind, the *counter-hypothesis*, H_1. Both hypotheses are assumed to be *simple*, in that each of them fully specifies the probability of the observation ω. It is convenient to assume that the experiment has only a discrete set of outcomes, so that ω can only take a discrete set of values, $\{\omega_k\}$. We shall use $P_i(\omega_k)$ to denote the probability of an outcome ω_k on hypothesis H_i.

Now, suppose the rule of inference is given the form that, if outcome ω_k is observed, H_0 will be rejected with probability x_k. The idea of a randomized decision may seem strange, but this formulation adapts the problem to our conditions, and yields, in the end, a sensible answer.

With this procedure, H_0 will be rejected with probability

$$\alpha_i = \sum_k x_k P_i(\omega_k) \tag{5.13}$$

if H_i is true. The x_k should be chosen so that α_0 is as small as possible, α_1 as large as possible. This question of simultaneous extremization of several functions is something we shall return to in Section (10.1). However, for the moment we shall adopt the criterion that α_1 is to be maximized subject to

$$\alpha_0 \leqslant \alpha, \tag{5.14}$$

where α is a pre-set small positive value, the *significance level*; an upper bound to the probability that one rejects a true hypothesis.

The problem is now reduced to that of maximizing α_1, a linear function of x, within

$$0 \leqslant x_k \leqslant 1 \tag{5.15}$$

subject to the linear constraint (5.14).

Suppose that the idea of randomized decisions had not been admitted, so that one demanded a predetermined decision one way or the other for each contingency ω_k. In this case the x_k would be restricted to taking the values 0 or 1, so that x would not vary in a convex domain, and expression (5.13) could not be regarded as a convex (or concave, as appropriate) function on $\mathscr{X}$. However, if randomized solutions are admitted, then the problem has the required convexity properties, and the solution satisfies, as in (5.11) and (5.12),

$$x_k = \begin{Bmatrix} 0 \\ \text{indeterminate} \\ 1 \end{Bmatrix} \quad \text{according as} \quad \lambda(\omega_k) \begin{Bmatrix} < \\ = \\ > \end{Bmatrix} y. \tag{5.16}$$

Here

$$\lambda(\omega) = \frac{P_1(\omega)}{P_0(\omega)}, \tag{5.17}$$

the *likelihood ratio*, and y is the non-negative scalar minimizing

$$F(y) + \alpha y = \sum_k [P_1(\omega_k) - y P_0(\omega_k)]_+ + \alpha y. \tag{5.18}$$

In all but rather degenerate cases (cf. Exercise (5.7)) $y > 0$, and equality holds in (5.14). The ω-set for which $\lambda(\omega) > y$ is termed the *rejection region*: if a value in this set is observed, H_0 is always rejected. The criterion makes sense: H_0 is rejected if the probability of ω on H_0 relative to that on H_1 falls below a certain level.

If ω is such that $\lambda(\omega) < y$, then H_0 is not rejected. In the transition region, the ω-set for which $\lambda(\omega) = y$ (which may have zero probability on one or both hypotheses), the rejection probability can take intermediate values, any assignment consistent with equality in (5.14) being acceptable. However, some assignments may be preferable on other grounds: see Exercise (5.9).

With only minor regularity conditions the same conclusions will hold when ω is distributed in a general space Ω, with a probability measure $P_i(\cdot)$ on sets of Ω if H_i holds ($i = 0, 1$). That is, the region R of definite rejection will be the ω-set for which

$$\lambda(\omega) \overset{\text{def}}{=} \frac{P_1(d\omega)}{P_0(d\omega)} > y, \tag{5.19}$$

where y is the non-negative scalar minimizing

$$F(y) + \alpha y = \int [\lambda(\omega) - y]_+ P_0(d\omega) + \alpha y. \tag{5.20}$$

If $y > 0$ and the set satisfying $\lambda(\omega) = y$ has zero P_0-measure, then this determination of y is equivalent to

$$P_0(R) = \alpha. \tag{5.21}$$

The question of hypothesis-testing (and of optimal decision-making, more generally) will recur in Section (10.7).

Exercises

(5.5) Relate the adapted simplex method outlined in the text to the dual problem, as in Section (4.7).

(5.6) Verify convexity of the function $F(y)$ of (5.12).

(5.7) Show that, in the hypothesis testing problem, y will be zero if and only if there is no set in Ω having positive probability on both hypotheses. That is, when an observation leads to complete and certain discrimination between the two hypotheses.

(5.8) Calculate the optimal rejection region in the case where the experiment consists of N independent tosses of a coin, with probability p of a head on each toss, and H_i specifies that p has the particular value π_i ($i = 0, 1$). That is, the probability of a particular sequence of results on H_i is $\pi_i^r(1 - \pi_i)^{N-r}$, where r is the number of heads in that sequence.

(5.9) Suppose that ω is the age at which a piece of equipment fails, and that the failure rate is zero up to some age a, and equal to ϕ thereafter. Thus, ω is a scalar with probability density

$$\rho(\omega) = \begin{cases} 0 & (\omega < a), \\ \phi e^{\phi(a-\omega)} & (\omega \geq a). \end{cases}$$

Suppose that a, ϕ have the values a_i, ϕ_i on H_i. Determine the optimal rejection region. Note that this is generally indeterminate in the case $\phi_0 = \phi_1$, when λ takes only two values (because the only relevant fact is whether the equipment fails before or after the larger of a_1, a_2: the actual time of failure, apart from this fact, does not help one to discriminate). However, the rational choice seems to be to take R as an interval, since this would be the limit solution as ϕ_0, ϕ_1 tend to equality from distinct values.

5.3 MAXIMAL FLOW IN A NETWORK

Some network problems were formulated in Section (4.4). Consider one particular example: the maximization of flow from one given node to another (the "source" and "sink") in a network carrying a single type of traffic, subject to capacity constraints on the arcs. This is the case when a network is required to carry as much material as possible from a single given origin point to a single given destination point. The cases of multiple traffic, origins and destinations are obviously more complicated. Sometimes there are also capacity constraints on the nodes, expressing handling limitations at intermediate depots. This feature will not be included, although it presents no formal problem: see Exercise (5.13).

Let the nodes of the network be labelled by $j = 1, 2, ..., n$ with source and sink corresponding to $j = 1$ and $j = n$ respectively. Let x_{jk} denote the rate of direct traffic flow from node j to node k. We distinguish between traffic in the two directions, at rates x_{jk} and x_{kj} respectively, and assume each non-negative:

$$x_{jk} \geqslant 0. \tag{5.22}$$

This represents a difference already between the traffic flow problem now considered, and the current flow problem of Exercise (1.11) and Section (3.7). In the latter case x_{jk} represented merely the *net* current flow from j to k, and could take either sign. In the present case, traffic in one direction does not cancel with that in another; both flows contribute to congestion of the channels, and they must be separately identified.

There are two different problems, according to whether the arcs are considered to be intrinsically directional (i.e. capable of taking traffic in only one direction) or not. We shall consider the directional case, in which distinction is made between arcs jk and kj; arc jk being capable of taking $j \rightarrow k$ traffic only, and being subject to a capacity constraint

$$x_{jk} \leqslant b_{jk}. \tag{5.23}$$

However, the treatment of this section adapts easily to the case of undirected arcs: see Exercise (5.12).

Let v denote the net flow through the network, from node 1 to node n. The balance equations

$$\sum_k (x_{jk} - x_{kj}) = r_j, \tag{5.24}$$

then hold, with

$$r_j = \begin{cases} v & (j=1), \\ -v & (j=n), \\ 0 & \text{(otherwise)}. \end{cases} \qquad (5.25)$$

The aim is to maximize v subject to the constraints (5.22)–(5.24). It is natural to take account of the conservation conditions (5.24) by use of Lagrangian multipliers; the question is, how should one take account of the capacity constraints (5.23)? If Lagrangian multipliers are associated with them also, the problem has LP structure: if, on the other hand,

$$0 \leqslant x_{jk} \leqslant b_{jk} \; (j,k = 1,2,...,n)$$

is regarded as the basic domain of variation $\mathcal{X}$, then the problem is one of allocation with bounded intensities. The second point of view seems preferable, since it keeps the number of multipliers to a reasonable figure, and since one would not normally wish to solve the problem for a range of capacity values b_{jk}.

If multipliers y_j are associated with constraints (5.24), then the Lagrangian form becomes

$$L(x,y) = v(1 + y_1 - y_n) - \sum_j \sum_k y_j(x_{jk} - x_{kj}). \qquad (5.26)$$

This is maximal when

$$\bar{x}_{jk} = \begin{cases} b_{jk} \\ \text{indeterminate} \\ 0 \end{cases} \quad \text{according as} \quad y_j \begin{cases} < \\ = \\ > \end{cases} y_k, \qquad (5.27)$$

and the dual problem is (cf. (5.12)): minimize

$$D(y) = \sum_j \sum_k b_{jk}(y_k - y_j)_+ \qquad (5.28)$$

subject to

$$y_n - y_1 = 1. \qquad (5.29)$$

Constraint (5.29) holds because v is unrestricted in range, so that $\max\limits_{x,v} L = +\infty$ unless v has zero coefficient in L.

This dual problem compares interestingly with the electrical network problem of Section (3.7). Again the flow x_{jk} is a function of $y_j - y_k$: the Lagrangian multipliers associated with conservation conditions at the nodes have the character of a potential. Relation (5.27) is "Ohm's law" for traffic flow, but an Ohm's law which is discontinuous rather than linear (see Figure (5.1)) because of the difference in the costing of flow. For the electrical case, the "cost" of carrying flow x_{jk} was proportional to x_{jk}^2. For our present case, constraint (5.23) implies that the cost is zero for $x_{jk} \leqslant b_{jk}$, infinite if $x_{jk} > b_{jk}$. The potentials y_j are now to be determined by minimization of expression (5.28) (subject to (5.29)), rather than of expression (3.61). For interpretation of these potentials, see Exercise (5.11).

Although in passing from primal to dual the number of variables has been decreased to n, one imagines that the minimization of expression (5.28) is

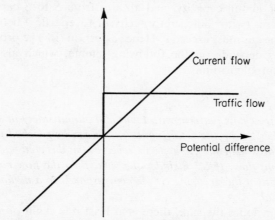

Figure 5.1 "Ohm's Law": the relationship of flow-rate to potential difference for flows of electric current and of traffic

comparable in difficulty with solution of the discrete Laplace equation (1.23). In fact, it is rather simpler, because it turns out that there is a solution of the dual problem for which all the y_j take one of the values 0 or 1. The proof of this fact is related to the idea of a "cut", to a beautiful theorem on flows and cuts due to Ford and Fulkerson, and to an algorithm for the primal, also due to these authors.

Suppose that the nodes of the network are grouped into two complementary sets: a set S containing the source, and its complement $\bar{S}$, containing the sink. The *cut* corresponding to this division is the set of arcs jk with j in S and k in $\bar{S}$. The *value of the cut* is the sum of the capacities of these arcs:

$$V = \sum_{j \in S} \sum_{k \in \bar{S}} b_{jk} \qquad (5.30)$$

Theorem (5.1)
    (i) *The value of a cut is not less than the value of the flow.*
    (ii) *The maximal flow equals the minimal cut.*

(iii) If $S, \bar{S}$ is the grouping of nodes corresponding to the minimal cut, then the dual problem has a solution

$$y_j = \begin{cases} 0 & (j \in S), \\ 1 & (j \in \bar{S}). \end{cases} \tag{5.31}$$

The value of the cut will equal the flow, v, if all arcs from S to $\bar{S}$ are carrying traffic at full capacity, and all arcs from $\bar{S}$ to S carry zero traffic. In all other cases (with some sub-capacity or reverse flow between S and $\bar{S}$) the value of the cut must exceed v. Hence assertion (i). The proof of the other two assertions depends on the following lemma, which also sketches the Ford–Fulkerson algorithm.

Lemma

Consider a trial flow pattern, with input and output solely at source and sink respectively. Let S^* be the set of nodes which can be reached from the source by a chain of arcs with sub-capacity or reverse (i.e. directed back to the source along the chain) flow. If S^* includes the sink, then the flow can be improved. If it does not, then the grouping $S^*, \bar{S}^*$ corresponds to a minimal cut, and the flow is maximal.

For, if S^* includes the sink, there is at least one chain of sub-capacity or reverse-flow arcs from source to sink. The direct flow along this chain can be increased until either direct flow has been brought up to the full capacity of some arc of the chain, or reverse flow is reduced to zero in some arc of the chain.

This chain is thus deleted, and iteration of the procedure must then ultimately bring one to the second case (see Exercise (5.10)) when S^* does not contain the sink and so defines a cut. Adding (5.24) over all j in S^*, one sees that

$$v = \sum_{j \in S^*} \sum_{k \in \bar{S}^*} (x_{jk} - x_{kj}), \tag{5.32}$$

the terms for k in S^* cancelling out in the summation. But $j \in S^*$, $k \in \bar{S}^*$ implies direct full-capacity flow from j to k, or node k would be in S^*. Hence $x_{jk} = b_{jk}$, $x_{kj} = 0$, and, from (5.32)

$$v = \sum_{j \in S^*} \sum_{k \in \bar{S}^*} b_{jk}, \tag{5.33}$$

the value of the cut associated with the grouping $S^*, \bar{S}^*$. By assertion (i), the flow must then be optimal, and the equality stated in assertion (ii) is proved.

One can now set $S^* = S$ without ambiguity. Now, suppose v a feasible net flow for the primal, U its maximal value, and y a vector obeying (5.29) and so feasible for the dual. Then

$$v \leqslant U \leqslant \sum_j \sum_k b_{jk}(y_k - y_j)_+,\qquad(5.34)$$

by the properties of the dual. For the y of (5.31) and the v of (5.33), equalities hold throughout in (5.34). The y of (5.31) thus solves the dual, as asserted in the theorem.

The Ford–Fulkerson algorithm is simply a systematic method of constructing the set S^* of the lemma, by working out iteratively from the origin along all sub-capacity and reverse flow arcs. If S^* contains the sink, then the flow is improved, as indicated above, and the whole procedure repeated. This seems to be the best algorithm, although the value of maximal v itself can be directly determined in those cases where the minimal cut is easily spotted. The minimal cut is the "bottleneck" of the network: see the last point in Exercise (5.11).

Exercises

(5.10) Suppose that capacities b_{jk} have integral values, and that one begins the Ford–Fulkerson computation with a trial flow with integral x_{jk}. Show that the improvement in flow is also integral, so long as improvement is possible, and hence that (a) the optimum is reached in a finite number of iterations, and (b) the x_{jk} are integral in the optimal solution.

(5.11) Note that only $n-1$ of the constraints (5.24) are functionally independent, and hence that the multipliers y_j are indeterminate up to an additive constant. Show that if the maximal flow U possesses derivatives then

$$y_k - y_j = \frac{\partial U}{\partial \delta},$$

where δ is the magnitude of a perturbation $r_j \rightarrow r_j - \delta$, $r_k \rightarrow r_k + \delta$. That is, if a by-pass of capacity δ from node j to node k were introduced, maximal flow could be increased by an amount $(y_k - y_j)\delta + o(\delta)$. Hence interpret solution (5.31).

(5.12) Suppose that there is a single arc of capacity b_{jk} between nodes j and k, which may carry traffic in either direction, so that condition (5.23) is modified to $x_{jk} + x_{kj} \leqslant b_{jk}$. Adapt the treatment of this section (the form of the dual, the theorem and the algorithm) to this case.

(5.13) Show that restrictions on node capacity can be represented by the introduction of dummy arcs, of restricted capacity.

(5.14) Consider the form of the dual in the case where jk traffic pays a toll of c_{jk}, so that one wishes to maximize $v - \sum \sum c_{jk} x_{jk}$.

(5.15) Consider primal and dual problems in the case where the network is a simple linear chain: $1 \rightarrow 2 \rightarrow 3 \rightarrow \ldots \rightarrow n$.

(5.16) The undirected arcs of the network shown in Figure (5.2) have the capacities indicated. By location of the minimal cut, or otherwise, determine a flow pattern which maximizes the total net flow from the node on the extreme left to that on the extreme right.

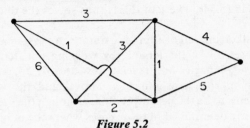

Figure 5.2

5.4 EXTREMAL DISTRIBUTIONS: TCHEBICHEV INEQUALITIES

Situations involving probability distributions in one way or another provide a fruitful source of linear problems. Indeed, if randomized solutions to an extremal problem are admitted, then all problems are thereby rendered linear.

For example, consider the problem $P_0(b)$: the maximization of $f(x)$ in $\mathcal{X}$ subject to $g(x) = b$. Suppose that x can only take a finite set of values in $\mathcal{X}$: then a randomized solution to $P_0(b)$ consists in determination of a distribution $\{p(x)\}$ on $\mathcal{X}$ which maximizes

$$E(f) = \sum_x p(x)f(x) \tag{5.35}$$

subject to

$$\sum_x p(x)g(x) = b. \tag{5.36}$$

That is, a set of constants $p(x)$ must be determined which maximizes the linear (in $p(x)$) expression (5.35) in the cone $p(x) \geqslant 0$ $(x \in \mathcal{X})$, subject to the linear constraints (5.36) and

$$\sum_x p(x) = 1. \tag{5.37}$$

Adoption of randomized solutions thus converts the problem $P_0(b)$ into an LP in the variables $p(x)$.

If $\mathcal{X}$ is a more general space, then one has to replace the idea of a discrete distribution $\{p(x)\}$ by that of a probability measure $\mu(\cdot)$ on appropriate sets of $\mathcal{X}$, and, to obtain a randomized solution to $P_0(b)$, one must determine the measure μ in some appropriately regular class which maximizes $\int f(x)\,\mu(dx)$, subject to

$$\int g(x)\,\mu(dx) = b. \tag{5.38}$$

This is still a problem with LP structure, but the fact that the variable is now taking values in a set of measures, possibly a rather general one, takes us into deeper waters than those we sailed in Chapter 4.

However, one does not in fact need to leave the haven of discrete distributions, for Theorem (3.13) assures us that a distribution providing a randomized solution to $P_0(b)$ can be found which is concentrated on at most $m+1$ points of $\mathcal{X}$. This is under the assumption that $g(x)$ is an m-vector, and that $(g(\mathcal{X}), f(\mathcal{X}))$ is finite, this last assumption being more one of convenience than of necessity.

A celebrated problem in probability concerns the calculation of *Tchebichev bounds*. Suppose that x is a random variable taking values in a set $\mathcal{X}$, and the expectation values $E[g_j(x)]$ of certain functions $g_j(x)$ of x are known. One asks: what are the greatest and least values that the expectation of another function, $f(x)$, can take, consistent with the given expectations? So far as the upper bound is concerned, then, the problem is: what is the distribution μ on $\mathcal{X}$ which maximizes $\int f(x)\,\mu(dx)$, subject to

$$\int g_j(x)\,\mu(dx) = b_j \quad (j = 1, 2, \dots, m)? \tag{5.39}$$

But this is nothing else than the problem of deriving a randomized solution to $P_0(b)$. One can consequently appeal to Theorem (3.13). In fact that theorem is now reproduced in a form more consistent with the literature on Tchebichev bounds.

Theorem (5.2)

Suppose that $f(x), g_1(x), g_2(x), \dots, g_m(x)$ are finite for x in $\mathcal{X}$, and that the vector $g(x) = (g_1(x), g_2(x), \dots, g_m(x))$ has prescribed expectation $E[g(x)] = b$, where b is interior to the convex hull of $g(\mathcal{X})$. Then

$$E[f(x)] \leqslant y_0 + \sum_1^m y_j b_j \tag{5.40}$$

for any set of coefficients $y_0, y_1, \dots, y_m$ consistent with

$$f(x) \leqslant y_0 + \sum_1^m y_j g_j(x) \quad (x \in \mathcal{X}) \tag{5.41}$$

and the sharp upper bound for $E(f)$ is obtained by minimizing the bound in (5.40) with respect to $y_0, y_1, \dots, y_m$ subject to (5.41). An extremal distribution,

*for which this bound is attained, is concentrated on the x-values for which
equality is then attained in (5.41), and an extremal distribution exists which is
concentrated on at most m+1 of these values.*

The theorem is, indeed, no more than a recasting of Theorem (3.13). If
y is used to denote the vector $(y_1, y_2, ..., y_m)$, as formerly, then the minimal
value of y_0 consistent with (5.41) is $\max_x [f(x) - y^T g(x)]$. The minimal value
of the bound in (5.40) is consequently $\min_y \max_x [f(x) + y^T(b - g(x))] = \hat{U}(b)$,
and it is known from Theorem (3.13) that this bound is attainable under the
conditions asserted, and with a distribution of the character claimed.

The assumption that f, g are finite in $\mathscr{X}$ is unrealistically restrictive. It is in
general easiest circumvented by a limit argument. Consider, for example, the
best-known simple example, the *Markov inequality* for the probability that a
non-negative scalar random variable with prescribed mean

$$E(x) = b \tag{5.42}$$

exceeds a value a. We have thus $g(x) = x$ and

$$f(x) = \begin{cases} 0 & (x < a), \\ 1 & (x \geqslant a), \end{cases} \tag{5.43}$$

the indicator function of the set $x \geqslant a$. In order to achieve finiteness, suppose
that x is restricted to the interval $0 \leqslant x \leqslant M$, which then constitutes the
domain $\mathscr{X}$. One may assume $M > a$.

It is desired to find coefficients y_0, u_1 which minimize $y_0 + y_1 b$ subject to

$$y_0 + y_1 x \geqslant f(x). \tag{5.44}$$

That is, it is desired to find the straight line $t = y_0 + y_1 x$ which lies above the
graph of the step function $t = f(x)$ for $x \geqslant 0$, and has minimal average
ordinate consistent with $E(x) = b$. The straight line which meets the graph
of $f(x)$ most closely from above would seem to be either $t = x/a$ (meeting
the graph of f at $0, a$) or $t = 1$ (meeting on $x \geqslant a$). These give the respective
bounds

$$\text{Prob}(x \geqslant a) \leqslant \begin{cases} b/a, \\ 1. \end{cases} \tag{5.45}$$

which are, in fact, sharp for the respective cases $a \geqslant b$, $a \leqslant b$ (see Exercise
(5.17)). The bound

$$\text{Prob}(x \geqslant a) \leqslant \min(b/a, 1) \tag{5.46}$$

is then sharp, and independent of M in $M > a$.

Exercises

(5.17) *The Markov inequality*. This is inequality (5.46). Prove its sharpness, by finding a distribution on the equality values in (5.44) for which the bound (5.46) is attained.

(5.18) Calculate a bound for $\text{Prob}(x \geqslant a)$ given that $M_1 \leqslant x \leqslant M_2$ and $E(x) = b$. Consider the effect on this bound of the transitions $M_1 \to -\infty$ or $M_2 \to +\infty$.

(5.19) *The Tchebichev inequality*. Suppose that x takes any real scalar value, with $E(x) = 0$, $E(x^2) = v$. Show that

$$\text{Prob}(|x| > a) \leqslant \min(v/a^2, 1)$$

and that this bound is sharp.

(5.20) Suppose that x takes values in the interval (α, β) and that the value of $b = E(x)$ is prescribed. Show that the following bounds are sharp:

$$E[(d-x)_+] \leqslant \begin{cases} 0 & (d \leqslant \alpha), \\ \dfrac{(\beta-b)(d-\alpha)}{\beta-\alpha} & (\alpha \leqslant d \leqslant \beta), \\ d-b & (d \geqslant \beta). \end{cases}$$

This result will be needed in Section (8.5).

(5.21) The example of the Markov inequality shows that strictly a a distinction should be made between a maximum and a supremum. Consider the case $b < a$, when the bound (5.46) is b/a. This provides a maximum for $\text{Prob}(x \geqslant a)$ (attained for a distribution concentrated at 0 and b/a) but merely a supremum for $\text{Prob}(x > a)$ (approached by a distribution concentrated on 0 and $(b/a + \varepsilon$ as $\varepsilon \downarrow 0)$. The effect is due to the lack of continuity of $f(x)$, as defined in (5.43).

5.5 THE CLASSIC TCHEBICHEV BOUND

The classic special case of the Tchebichev inequality is that in which one sets bounds on the distribution function of a scalar variable x from a knowledge of some of its moments, $E(x^j)$. Explicitly, x takes values in a set $\mathcal{X}$ on the real line; the moments

$$b_j = E(x^j) \tag{5.47}$$

are given for a number of values of j, and it is desired to set bounds on the value of the distribution function

$$F(a) = \text{Prob}(x \leqslant a) \tag{5.48}$$

at some particular point a.

The problem is mainly one of historical interest, for it was proposed originally as a step towards proof of the central limit theorem; a topic now treated by quite other methods. However, the solution (due to Tchebichev) is so attractive as to deserve reproduction. Moreover, it affords a good example of the analytic solution of a sufficiently structured linear problem.

The moments b_j will be assumed given for $j = 0, 1, 2, ..., m$ (with $b_0 = 1$) and only the case where $\mathscr{X}$ is the entire real axis, R^1 will be considered. The distribution function can be written

$$F(a) = E[f(x)], \tag{5.49}$$

where

$$f(x) = \begin{cases} 1 & (x \leqslant a), \\ 0 & (x > a), \end{cases} \tag{5.50}$$

the indicator function of the set $x \leqslant a$.

Converting the problem to dual form one must then find polynomials in x of degree m which bound $f(x)$ from above and below as closely as possible (in the sense that the expectation of the polynomial is respectively minimal and maximal). In Figure (5.3) approximation from above is illustrated in the

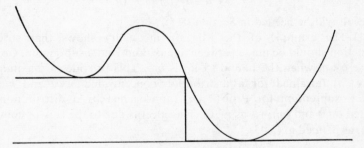

Figure 5.3 A polynomial bounding a simple step function as closely as possible from above

case $m = 4$. Such polynomial approximations have a special theory of their own (see Karlin and Studden, 1966) but we shall work from first principles.

An appeal to Theorem (5.2), if valid, would imply that the extremal distribution must be concentrated on the values of x for which the bounding polynomial equals $f(x)$. Now, since in the case $\mathscr{X} = R^1$ the bounding polynomial must be of even degree, in both the cases $m = 2r, 2r+1$ it must be of degree $2r$ at most. As follows from the figure, there will then be at most $r+1$ contact points: at $x = a$ and at the r (at most) turning points of the polynomial. If $\mathscr{X}$ is not the whole axis then the situation is different; bounding polynomials of odd degree are possible, and the end-points of intervals in $\mathscr{X}$ may provide additional contact points.

Theorem (5.3)

The bounds which can be set on $F(a)$ in the cases $m = 2r, 2r+1$ (r integral) are identical. If $b_0, b_1, ..., b_{2r}$ are consistently prescribed, then a distribution on at most $r+1$ points (one of these being $x = a$, if $r+1$ points are needed) with these moments exists. If this has distribution function $G(x)$, then G is uniquely determined at its continuity points, and

$$G(a-) \leqslant F(a) \leqslant G(a+). \qquad (5.51)$$

These bounds on F are sharp.

The proof of the theorem provides a construction for G.

The discussion before the theorem showed that, in both the cases $m = 2r$ and $m = 2r+1$, the extremal distribution must be concentrated on at most $r+1$ points, one of these certainly being $x = a$ if $r+1$ points are needed. It is implied that such a distribution with the given moments exists. If it can be shown that this is uniquely determined at its continuity points, then the statement that bounds (5.51) hold is immediate.

However, these conclusions follow only by appeal to Theorem (5.2), and two of the regularity conditions of that theorem are not fulfilled. The functions $g_j(x) = x^j$ are not finite in R^1, and it is not necessarily assumed that the given moment values are interior to the set of possible moment values. The conditions of Theorem (5.2) can be relaxed to admit these contingencies, at least in the case $m = 2r$. However, in the present case it will follow directly from the construction that the distribution exists and provides the sharp bounds.

Consider an s-point distribution which assigns probabilities p_k to distinct values x_k. (So x_k denotes a particular value of x, rather than a scalar component of a vector x, in this section.) If this has the prescribed moments, then

$$\sum_1^s p_k T(x_k) = E[T(x)] \qquad (5.52)$$

for any polynomial $T(x)$ of degree $2r$ or less. The expectation $E[T(x)]$ is one consistent with the prescribed moments. Denote the polynomial with roots x_k by $Q(x)$:

$$Q(x) = \text{const} \prod_1^s (x - x_k). \qquad (5.53)$$

Now, in the degenerate case $s \leqslant r$ (which is just that in which $b_1, b_2, ..., b_{2r}$ is a boundary point of the permitted set of moment values) one can set $T = Q^2$, and so deduce from (5.52) that

$$E[Q(x)]^2 = 0. \tag{5.54}$$

That is, *any* distribution consistent with the given moments is concentrated on the zeros x_k of Q. Since the x_k are distinct, relation (5.52) determines the weights uniquely. In fact, setting

$$T(x) = [U_k(x)]^2 \overset{\text{def}}{=} \prod_{j \neq k} \left(\frac{x - x_j}{x_k - x_j} \right)^2, \tag{5.55}$$

one sees from (5.52) that

$$p_k = E[U_k(x)]^2. \tag{5.56}$$

The s-point distribution G is thus uniquely determined, and it follows from (5.54) and (5.56) that any distribution function $F(x)$ consistent with the given moments can differ only from $G(x)$ by an infinitesimal variation of the values x_k. Bounds (5.51) hence follow in the degenerate case.

The polynomial $Q(x)$, and hence the values x_k, can be determined from the given moments b_j by choosing $T(x) = x^j Q(x)$ in (5.52), whence

$$E[x^j Q(x)] = 0 \quad (j = 0, 1, \ldots, 2r - s). \tag{5.57}$$

The first s of these equations determine Q. (See the argument for the non-degenerate case, below.)

In the non-degenerate case s must certainly equal $r + 1$ if (5.52) is to be satisfied. The system (5.57) still provides r distinct equations for the determination of Q and the x_k: the $(r + 1)$th follows from

$$Q(a) = 0. \tag{5.58}$$

Relations (5.57) and (5.58) determine Q uniquely (to within a multiplicative factor). If they did not, then a Q of lower order than $r + 1$ could be found to satisfy them, and one would again be in the degenerate case. Furthermore, the $r + 1$ roots x_k determined by (5.57) and (5.58) must be real and distinct. If they were not, then $Q(x)$ would show at most $r - 1$ changes of sign as x varied (since complex roots occur in conjugate pairs, and repeated roots at least in pairs). A polynomial $R(x)$ could then be found which changed sign with Q, so that $QR \geqslant 0$, and yet of at most degree $r - 1$, so that

$$E(QR) = 0. \tag{5.59}$$

This would imply a distribution concentrated on the zeros of $R(x)$, and one would again be in the degenerate case.

Thus (5.57) and (5.58) imply a construction for and a unique determination of $Q(x)$ and the x_k: (5.55) and (5.56) a construction for and a unique determination of the p_k. This uniqueness would prove the theorem, if appeal to Theorem (5.2) were valid. However, appeal is not necessary. The $(r+1)$-point distribution $G(x)$ thus determined has the prescribed moments b_j up to order $2r$, so that (5.52) becomes

$$\int T(x)\,dG(x) = \int T(x)\,dF(x) \qquad (5.60)$$

for any distribution function F consistent with the prescribed moments. That is,

$$\int T'(x)\,[F(x)-G(x)]\,dx = 0. \qquad (5.61)$$

Now, $T'(x)$ is an arbitrary polynomial of degree $2r-1$. It consequently follows from (5.61), as in the arguments associated with (5.54) and (5.59), that either $F \equiv G$ or $F - G$ changes sign strictly at least $2r-1$ times. That is, the graphs of F and G must interlace as in Figure (5.4). Bounds (5.51) are

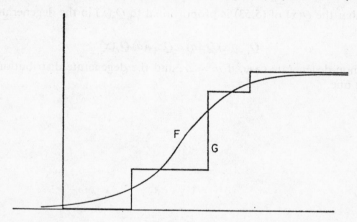

Figure 5.4 The interlacing of two distribution functions whose first four moments coincide. Typical graphs of F and G in the case $r = 2$, when G has $r+1 = 3$ steps and there are $2r-1 = 5$ crossing points

hence valid, and they can be realized by the extreme cases $F(x) = G(x-)$, $G(x+)$.

The reason why the bounds for the case $m = 2r+1$ are no better than for the case $m = 2r$ can be loosely expressed as follows. Suppose that a distribution is modified by removal of a probability mass δ from some fixed co-ordinate to co-ordinate $(|K|/\delta)^{1/(2r+1)}\,\mathrm{sgn}\,K$. The values of F and

b_j $(j < 2r+1)$ are thereby modified by an amount which tends to zero with δ, but b_{2r+1} is modified by $K + o(\delta)$. Thus b_{2r+1} can take any value consistent with a given F, and prescription of its value does not constrain F (in the case $\mathscr{X} = R^1$).

If $\mathscr{X}$ is a finite interval, it is found (in the case $m = 2r+1$) that the bounding polynomial is in general of full degree, $2r+1$, meeting $f(x)$ also at one end of the interval. As the interval becomes infinite, the term of highest degree tends to zero, but there still is a contact point at infinity. The probability weight at infinity can be vanishing, but still affect b_{2r+1}, as has just been seen.

Exercise

(5.22) A sequence of orthogonal polynomials $\{Q_r(x)\}$ can be associated with any distribution possessing moments, such that $Q_r(x)$ is of degree r exactly, and

$$E[(Q_r(x)\,Q_s(x)] = 0 \quad (r \neq s;\ r, s = 0, 1, 2, \ldots).$$

Show that the $Q(x)$ of (5.53) is proportional to $Q_s(x)$ in the degenerate case, and to

$$Q_{r+1}(x)\,Q_r(a) - Q_{r+1}(a)\,Q_r(x)$$

in the non-degenerate case, if $m = 2r$, and the degenerate distribution is an s-point one.

CHAPTER 6

Some Problems with Linear Constraints. I

It is natural to give the linear constraint special attention. For one thing, it is as yet only for the case of linear $g(x)$ that strong Lagrangian methods can be applied to a situation with equality constraints; Theorem (3.10). For another, in the case of linear $g(x)$, relatively mild additional hypotheses will exclude the possibility of infinite gradients of $U(b)$; Theorem (2.3).

As mentioned in Section (3.6), the special feature of the linear case is that supporting hyperplanes to the graph of $U(b)$ are immediately related to supporting hyperplanes to the graph of $f(x)$.

So many practical problems are of the linear constraint type that both this and the next chapter are devoted to examples. The individual sections of the two chapters are largely independent. The lengthier and more specialized treatments (such as that of structures in Sections (7.2) and (7.3)) have mainly been collected in Chapter 7.

6.1. ALLOCATION WITH NONLINEAR RETURN

Consider the allocation problem of relations (4.1)–(4.3), with the modification that the return function $f(x)$ need not be linear. That is, it is desired to maximize $f(x)$ in $x \geqslant 0$ subject to

$$Ax \leqslant b. \qquad (6.1)$$

Suppose $f(x)$ concave. Then an appeal to Theorem (3.9) shows that the solution maximizes some form $wf(x) - y^{\mathrm{T}} Ax$ in $x \geqslant 0$, with $w, y \geqslant 0$. If b is interior to $\mathscr{B}$ and U finite in a neighbourhood of b, then w can be set equal to unity, and y has the now familiar price interpretation. The dual problem will, however, not have the same symmetric relation to the primal as it did in the wholly linear case of Sections (4.1) and (4.3).

Nor is there in general any statement corresponding to the assertion of Theorem (4.2) that the optimal solution can be taken as basic, i.e. as having the minimum feasible number of non-zero intensities x_k. Indeed, strict concavity of $f(x)$ will encourage a spreading of investment. For example, consider the simplest case, where f is of the additive form

$$f(x) = \sum_k f_k(x_k), \tag{6.2}$$

and there is a single constraint

$$\sum x_k \leqslant b.$$

Then, in the case of concave f_k there exists a non-negative scalar y such that x_k maximizes $f_k(x_k) - y x_k$ in $x_k \geqslant 0$ $(k = 1, 2, ..., n)$. x_k will thus be greater than zero for all k such that $f_k'(0) > y$, and this could conceivably be true for any number of the k's.

The situation is illustrated in Figure (6.1). In case (i) the overall profit rate y is attained for a positive value of x_k for activity k, which consequently gives the optimal level of allocation $\bar{x}_k$ for this activity. In case (ii) the rate y is attained for no x_k; the activity is consequently intrinsically unprofitable, and $\bar{x}_k = 0$.

If f is non-concave, then the strong Lagrangian method will still supply an optimal solution, although conceivably a randomized one. In case (iii) the critical rate y is attained at two values, x_k' and x_k''. The value of x_k adopted will be that out of 0, x_k' and x_k'' which maximizes $f_k(x_k) - y x_k$, if the maximum is reached at only one of these points. If the maximum is attained at more than one of these values (as in cases (iv) or (v)) then all such values may enter the solution, in that they may be assigned a positive probability of adoption.

The specification of a nonlinear utility $f(x)$ is particularly necessary for the model of an optimal equilibrium economy of Section (5.1). In this context the constraint is of the form (6.1) with A a difference of non-negative matrices, so that the elements of A can take either sign. In such a case there may be feasible x with infinite elements, which may well yield an infinite utility if f is linear. However, the point is that f is not linear: an unlimited supply of plastic gnomes would not compensate at all for a lack of potatoes, and an infinite supply of turnips would only provide a partial substitute.

A reasonable first attempt at representing a situation in which there is no substitution between products (or activities) might be to set

$$f(x) = \min_k (x_k/d_k). \tag{6.3}$$

That is, the intrinsic utility of activity k is measured on a scaling x_k/d_k, and the value of the combined utility is dictated by the smallest component utility. The product in shortest supply, on an appropriate scale, is the critical one. The smaller d_k, the less necessary is activity k, in that the smaller x_k need

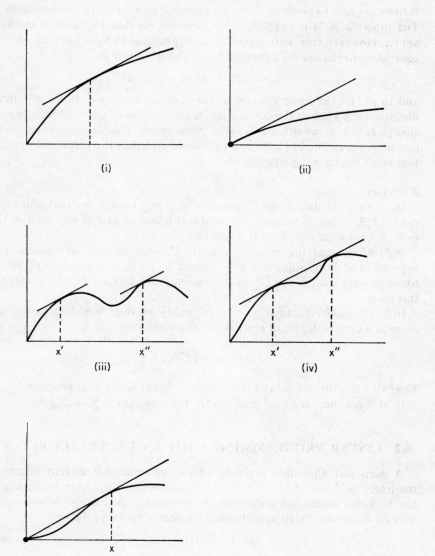

Figure 6.1 The determination of optimal intensity for one activity of an allocation problem with nonlinear return

then be in order to bring x_k/d_k above a prescribed value of the minimum (6.3). The function (6.3) is nonlinear, but concave, so that Lagrangian methods apply. However, the maximization of expression (6.3) subject to (6.1) is equivalent to the maximization of the scalar u subject to

$$x_k - ud_k \geqslant 0 \quad (k = 1, 2, ..., n) \tag{6.4}$$

and to (6.1). (The basic domain of variation is, of course, still $x \geqslant 0$, and u unrestricted.) That is, the constrained maximization of the nonlinear expression (6.3) can be rewritten as a linear programme. This device is often useful (see Sections (8.3) and (8.5)) and in a sense underlies the successful application of strong Lagrangian methods.

Exercises

(6.1) Suppose that unused resources of type j have a residual unit value v_j $(j = 1, 2, ..., m)$. Determine the modified forms of primal and dual in this case. Within what region must y now lie?

(6.2) Show that the problem dual to the constrained maximization of expression (6.3) is: minimize $y^T b$ subject to $y \geqslant 0$, $y^T A \geqslant 0$ and $y^T Ad \geqslant 1$. Show that the dual of the LP reformulation in the text can be transformed to this form.

(6.3) Consider the modification of utility (6.4) in which activities are allowed a certain degree of mutual substitutability, so that

$$f(x) = \min_i \left(\sum_k h_{jk} x_k \right)$$

Find an LP reformulation of this problem. What is the dual problem?

(6.4) Formulate the dual problem for the case $f(x) = \sum_k c_k \log x_k$.

6.2   LINEAR PROGRAMMING WITH AN UNCERTAIN RETURN

A particular allocation problem which has some independent interest is the linear problem of (4.1)–(4.3), in the case where the return coefficients c are to some extent unpredictable. A reasonable evaluation of the utility derived from an allocation with intensity vector x might now be

$$f(x) = E[H(c^T x)]. \tag{6.5}$$

Here E is the expectation operator, averaging over variable c, and H a concave function. The concavity of H gives a premium to low variability of return.

The problem is again one of maximizing $f(x)$ in $x \geqslant 0$ subject to $Ax \leqslant b$, and the Lagrangian theory of the previous section will apply. Since $f(x)$ is now nonlinear, the solution to the problem will no longer be basic, in general.

Uncertainty thus has the effect of causing a spread of investment. The rather extreme strategy of the wholly linear problem is modified in the direction of caution.

As a first approximation to the stochastic case, suppose that H can be expanded in powers of deviations of $c^T x$ from its mean value, the contribution from powers higher than the second being negligible.

Let the expected value and the variance of the actual return $c^T x$ be denoted by μ and v respectively. Then

$$f(x) = E[H(c^T x)] \sim H(\mu) + \tfrac{1}{2}vH''(\mu), \qquad (6.6)$$

the prime indicating differentiation, and

$$H^{-1}(f(x)) \sim \mu + \frac{H''(\mu)}{2H'(\mu)} v. \qquad (6.7)$$

If H^{-1} is monotone increasing in the effective range of $f(x)$, then maximization of $f(x)$ is equivalent to maximization of $H^{-1}(f(x))$ or of

$$f^*(x) = \mu - \tfrac{1}{2}\theta v = \sum_k \bar{c}_k x_k - \tfrac{1}{2}\theta \sum_j \sum_k v_{jk} x_j x_k, \qquad (6.8)$$

where $\bar{c}_k$ is the expected value of c_k, v_{jk} the covariance of c_j and c_k, and θ equals $-\tfrac{1}{2}H''(\mu)/H'(\mu)$. If this last ratio is effectively independent of μ (as will be exactly true if H is exponential, or approximately true if H'' is small and constant) then the expression (6.8) to be maximized is a quadratic function of x, reducing to the linear form $c^T x$ in the case of constant c. The coefficient θ measures the loss in return caused by uncertainty, or variability.

Consider, for explicitness, the special case where the c_k are uncorrelated:

$$f^*(x) = \sum \bar{c}_k x_k - \tfrac{1}{2}\theta \sum v_{kk} x_k^2, \qquad (6.9)$$

and there is a single constraint

$$\sum a_k x_k \leqslant b, \qquad (6.10)$$

all coefficients being positive. In the deterministic case ($v_{kk} \equiv 0$) the allocation is concentrated on those values of k maximizing $\bar{c}_k/a_k$. If y is the multiplier associated with (6.10), then the optimal allocation will be of the form

$$\bar{x}_k = \begin{cases} \dfrac{(\bar{c}_k - ya_k)}{\theta v_{kk}} & (\bar{c}_k \geqslant ya_k), \\[2mm] 0 & (\bar{c}_k \leqslant ya_k), \end{cases} \qquad (6.11)$$

or $\bar{x}_k = (\bar{c}_k - ya_k)_+/\theta v_{kk}$, where y is the non-negative value minimizing

$$\frac{1}{2\theta} \sum_k (\bar{c}_k - ya_k)_+/v_{kk} + by. \qquad (6.12)$$

It is seen from (6.11) that investment is more widely spread than in the deterministic case. Note, also, that if the stationarity condition for expression (6.12)

$$\sum_k \frac{a_k(\bar{c}_k - ya_k)_+}{\theta v_{kk}} = b \tag{6.13}$$

does not have a non-negative root, then $y = 0$, and investment does not reach the limit b. This is the situation in which the relative variability of return is so large that to invest all resources would be to court loss.

6.3 THE REGULATION OF A DAM WITH A DETERMINISTIC INFLOW

An interesting problem is that of determining how the rate of release of water from a storage reservoir, such as a dam, should be programmed in order to maximize utility derived over a period of time. This is a dynamic problem, essentially one of control, but one that can usefully be reformulated as a constrained maximization problem. Dynamic situations have already been considered in this way in Sections (2.5) and (5.1), and will be again in Sections (7.1) and (8.2).

Consider the amount of water in the dam, x_t, at consecutive instants of time $t = 0, 1, 2, ..., N$. If the dam has capacity C then x is restricted by

$$0 \leqslant x_t \leqslant C. \tag{6.14}$$

The sequence $\{x_t\}$ will obey the sequence of balance equations

$$x_{t+1} = x_t + v_t - u_t, \tag{6.15}$$

where v_t is the inflow between instants t and $t+1$ and u_t the corresponding outflow, or draw-off. The sequence $\{v_t\}$ will be assumed positive, prescribed and known. On the other hand, the outflow sequence $\{u_t\}$ can be chosen arbitrarily, subject to (6.14) and

$$u_t \geqslant 0. \tag{6.16}$$

It will be assumed that $u_0, u_1, ..., u_{N-1}$ are chosen so as to maximize

$$f = \sum_0^{N-1} h(u_t), \tag{6.17}$$

where $h(u)$ is a non-decreasing concave function, representing the utility of an outflow u (for irrigation, power or urban use, for example).

The problem can be regarded as one of constrained maximization, in which f is to be maximized with respect to $(x_1, x_2, ..., x_N; u_0, u_1, ..., u_{N-1})$. The basic region of variation is described by (6.14) and (6.16) and prescription of x_0 (although if h were chosen so that $h(u) = -\infty$ for $u < 0$, then the explicit restriction (6.16) could be dropped). The dynamic relation (6.15) will be regarded as a sequence of constraints, to be taken into account by Lagrangian methods. The concavity of the utility function $h(u)$ implies that, for a given value of total outflow $\sum_0^{N-1} u_t$, a sequence $\{u_t\}$ which is relatively constant will be preferred to one that is not. This is in fact the role of the dam: to act as a smoothing device, which produces a relatively smooth output $\{u_t\}$ from a variable input $\{v_t\}$. Unfortunately, it is a device which saturates: as soon as either of the bounds of emptiness or fullness is reached in (6.14), smoothing of decreasing or increasing flow (respectively) becomes impossible. The larger the capacity of the dam, C, the less often will this happen, and the greater the capacity of the dam to buffer variations in inflow.

The constraints (6.15) can be written in the form

$$x_{t+1} - x_t + u_t = v_t \quad (t = 0, 1, ..., N-1), \qquad (6.18)$$

corresponding to the form $g(x) = b$ of the general constraint. The inflows v_t correspond to the elements of b, the "resource availabilities"; by varying these one generates the whole family of constrained problems. The set $\mathscr{B}$ is the set of $(v_0, v_1, ..., v_{N-1})$ representable as in (6.18) with x_0 prescribed, and remaining x_t, u_t constrained by (6.14) and (6.16). The only boundary points of $\mathscr{B}$ which lie in $v \geqslant 0$ are those vectors for which $v_0 = 0$ if also $x_0 = 0$. So all non-negative vectors v but these are strictly interior to $\mathscr{B}$.

If h is concave, then all convexity conditions of Theorem (3.10) are satisfied, and conditions (6.18) (or (6.15)) can be accounted for by a multiplier vector y. The element y_t is $\partial U/\partial v_t$ (in the literal or some generalized sense) and so represents a price for water supplied at time t. The only question is whether these prices are finite (i.e. whether w can be set equal to unity in Theorem (3.9)). Both N and the v_t are supposed finite. If $h(u)$ is bounded for non-negative finite u, then so is U for finite v, since

$$Nh(0) \leqslant U(v) \leqslant Nh\left(\frac{1}{N} \sum_0^{N-1} v_t\right). \qquad (6.19)$$

Thus y will be finite if $(v_0, v_1, ..., v_{N-1})$ is interior to its permitted set (by appeal to (iiib) of Theorem (3.9)), and this will be so unless $x_0 = v_0 = 0$. y can be taken as finite even in this latter case if U has finite directional derivatives in v (see condition (iiic) of the same theorem), which is readily seen to be so if $h(u)$ has bounded derivative in $u \geqslant 0$.

Granted these regularity conditions on h, one can then find the solution by maximizing the Lagrangian form

$$L(x, u; y) = \sum_{t=0}^{N-1} [h(u_t) - y_t(x_{t+1} - x_t - v_t + u_t)] \qquad (6.20)$$

with respect to x, u. The variable x_t enters linearly, with coefficient $y_t - y_{t-1}$, so that then

$$\left.\begin{matrix} x_t = 0 \\ 0 \leqslant x_t \leqslant C \\ x_t = C \end{matrix}\right\} \quad \text{according as} \quad y_t \left\{\begin{matrix} < \\ = \\ > \end{matrix}\right\} y_{t-1}$$

or

$$y_t \left\{\begin{matrix} \leqslant \\ = \\ \geqslant \end{matrix}\right\} y_{t-1} \quad \text{according as} \quad \left.\begin{matrix} x_t = 0 \\ 0 < x_t < C \\ x_t = C \end{matrix}\right\} \quad (t = 1, 2, \ldots, N-1). \quad (6.21)$$

The three cases in (6.21) will be referred to as the dam's being in an *empty*, *intermediate* or *full* phase, respectively. Relations (6.21) will also hold for $t = N$ if y_N is assigned the value zero. (That is, water has no terminal value.) In the case of indefinite continuation, with a period of length N, $y_N = y_0$ would be the appropriate boundary condition.

Since L is also maximal with respect to u_t, then u_t must maximize $h(u) - y_t u$. Collect now the assertions that can be made;

(i) *The outflow u_t maximizes $h(u) - y_t u$.*

Note that if the u maximizing $h(u) - yu$ is denoted by $u(y)$, then the function $u(y)$ is decreasing; see Figure (6.2).

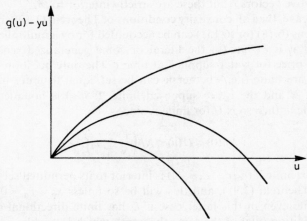

Figure 6.2 Graphs of $g(u) - yu$ for varying y. The curve drops as y increases

(ii) If $0 < x_t < C$ then $y_t = y_{t-1}$, so that $u_t = u_{t-1}$. Thus, *the outflow u_t is constant in an intermediate phase* (although in general having different values in distinct intermediate phases).

(iii) If $x_t = 0$ then $y_t \leqslant y_{t-1}$, so that $u_t \geqslant u_{t-1}$. Since $u_t \leqslant v_t$ and $u_{t-1} \geqslant v_{t-1}$ if neither x_{t+1} nor x_{t-1} is to be negative, it also follows that $v_t \geqslant v_{t-1}$. That is, *inflow is increasing in an empty phase.*

(iv) Correspondingly, *inflow is decreasing in a full phase.* Explicitly, if $x_t = C$, then $y_t \geqslant y_{t-1}$ and $u_t \leqslant u_{t-1}$, $v_t \leqslant v_{t-1}$.

(v) *If $h(u)$ is strictly increasing, then $y_t > 0$ ($t = 0, 1, ..., N-1$) and $x_N = 0$.* For it follows from (i) that, if $y_t \leqslant 0$, then $u_t = +\infty$, which cannot happen. Thus $y_{N-1} > 0$, but since $y_N = 0$ it follows from (6.21) that $x_N = 0$. Essentially, the fact that marginal utility $h'(u)$ is always positive implies that the "shadow price" for water is always positive, and that all water will be used by the end of the period.

The natural question is now, whether the above conditions are sufficient to determine the optimal policy for release of water. In fact, they are not. At least one further condition is required, which is simpler to specify in the case of continuous than of discrete time. For this reason, the continuous time version of the problem will now be considered, with $x(t)$ denoting the amount of water in the dam at time t, and $u(t)$ and $v(t)$ the outflow and inflow *rates*. We have then, corresponding to (6.14) and (6.15)

$$0 \leqslant x(t) \leqslant C, \tag{6.22}$$

$$\dot{x} = v - u, \tag{6.23}$$

where $\dot{x}$ is the time rate-of-change of x. It is supposed that the object is now to choose $u(t)$ so as to maximize $\int_0^N h(u(t))\, dt$.

Theorem (6.1)

Suppose that $h(u)$ is strictly concave, finite for finite positive u, that $v(t)$ is continuous and that x, v are not both initially zero. Then

(i) *u is constant in an intermediate phase.*

(ii) *If $x = 0$ then $u = v$, and v is increasing.*

(iii) *If $x = C$ then $u = v$, and v is decreasing.*

(iv) *u is continuous at a change of phase, except possibly at $t = N$.*

(v) *If $h(u)$ is strictly increasing, then $x(N) = 0$.*

With one exception, these assertions are direct analogues of results already established in the discrete case, and it will be assumed that they carry over from that case. (To prove the point requires a treatment of constrained maximization in function space.) The exception is the vital assertion (iv), which gives the transition rule at a change of phase. Assertion (ii) of the discrete case states that $u_t = u_{t-1}$ if the dam is in an intermediate phase at time t, which implies, in the continuous case, that $u(t)$ will be continuous when the dam *enters* an intermediate phase from a full or empty phase. It is now desired to establish continuity at *exit* from an intermediate phase. This continuity condition at exit seems to have no simple analogue in the discrete time case, and requires special proof.

Consider an intermediate/empty transition (see Figure 6.3). Suppose that the dam contains amount x at time t_0, and that it becomes empty at time τ. If

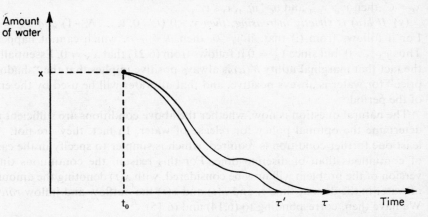

Figure 6.3 The variation of passage to an empty phase with variation in release rate

between t_0 and τ the water is being released at rate u (necessarily constant) then we must have

$$\int_{t_0}^{\tau} (u - v(t))) \, dt = x. \tag{6.24}$$

If u were changed, then so would τ be, and it would follow from (6.24) that

$$\tau - t_0 + (u - v(\tau)) \frac{\partial \tau}{\partial u} = 0. \tag{6.25}$$

Thus, if u is changed to $u' = u + \delta$, then τ is changed to

$$\tau' = \tau - \Delta = \tau - \frac{(\tau - t_0)\,\delta}{u - v(\tau)}, \tag{6.26}$$

to first order in δ. The change in utility is

$$\int_{t_0}^{\tau'} h(u') \, dt - \int_{t_0}^{\tau} h(u) \, dt + \int_{\tau'}^{\tau} h(v) \, dt \sim h'(u)\,(\tau - t_0)\,\delta + (h(v(\tau)) - h(u))\,\Delta$$

$$\sim (\tau - t_0)\,\delta \left[h'(u) - \frac{h(u) - h(v(\tau))}{u - v(\tau)} \right]. \tag{6.27}$$

However, the maximal utility must be stationary with respect to variations in u, so that the square bracket in (6.27) must be zero. Since h is strictly concave, the equality

$$h'(u) = \frac{h(u) - h(v(\tau))}{u - v(\tau)}$$

is possible only if $u = v(\tau)$. Since $v(\tau)$ is certainly the value adopted by u immediately *after* entry into the empty phase, one sees that u is indeed continuous in the transition.

The reason why this argument has no simple analogue in the discrete time case is that the moment of transition cannot then be varied continuously, as was possible here. The point $t = N$ is excepted in (iv), for the same reason.

These conditions will in most cases determine the policy completely, and in any case narrow it down to a small number of alternatives. Starting from (x, t) a release rate u is chosen which ensures that either (a) the dam becomes empty at time N, without previous change of phase, or, (b) $u = v(\tau)$ at the moment τ of first change of phase. In the rare event that there are several possible τ, the latest will presumably be chosen, but this is a matter for individual test. The same rules guide the decision of when to leave a full or empty phase (u being continuous on exit), with the additional provisos that rules (ii) and (iii) must be observed.

Some particular cases are drawn in Figures (6.4) and (6.5) on the assumption of a periodic v (i.e. a seasonal pattern) and the assumption that an equilibrium policy has been reached in which u and x are also periodic. The condition of periodicity thus replaces the terminal condition $x(N) = 0$ in this case.

In the case of Figure (6.4) there are one empty, one full and two intermediate phases in each cycle. The transition points are fully determined by continuity of u and by the requirement that x change by $\pm C$ over each intermediate phase (i.e. that the shaded regions in Figure (6.4) each have area C). It is seen that the effect of the dam is to slice peaks and troughs off the graph of $v(t)$, removing compensating areas equal to C.

If C is so large that one cannot remove areas as in Figure (6.4) without the consequent values of u crossing, then the situation is that of Figure (6.5). Here the dam is so large that it can buffer all v variation. The dam can be held in an intermediate phase, and u constant, throughout the season. In this case the shaded areas above and below the u line are equal, but less than C in value.

Suppose that v shows the minor variations of Figure (6.6). Because of rule (iii), it is not optimal to hold $x = C$ and $u = v$ throughout the final local maximum and minimum on the v curve: at some stage the water level is

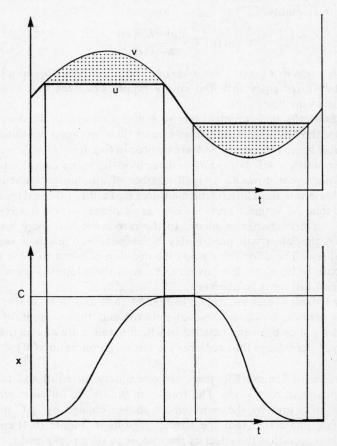

Figure 6.4 A diagram of inflow rate (v), release rate (u)
and water level (x) for a situation periodic in time. The
dam is brought from the empty state to the full state, at
constant release rate, during an interval of maximum
inflow, the reverse occurs during an interval of minimal
inflow. The effect is that the graph of u is derived from
that of v by removal of compensating peaks and troughs,
each area being equal to the dam capacity, C

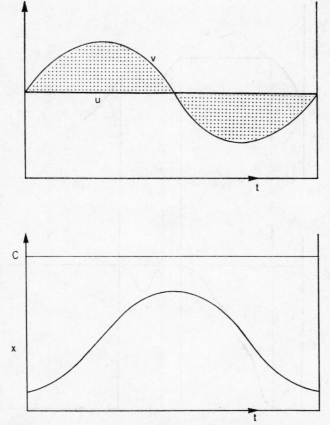

Figure 6.5 As in Figure (6.4), except that the dam is
sufficiently large relative to inflow fluctuations that it
becomes neither empty nor full throughout the period,
and a constant release rate can be maintained

allowed to fall, and then rise again. The moment of transition is determined
by continuity of u, and equality of the two shaded areas. The effect of the
manœuvre is that these minor variations are smoothed out, on a constant u
value.

Exercises

(6.5) Suppose the dam subject to evaporation and leakage, so that
equation (6.15) must be modified to

$$x_{t+1} = \rho x_t + v_t - u_t$$

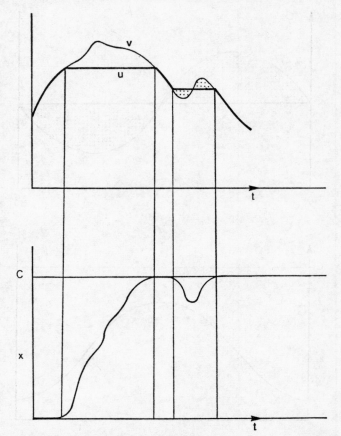

Figure 6.6 An illustration of how a local variation of input flow will be smoothed out, by maintenance of a constant release rate. The moment at which one leaves the full for the intermediate phase is determined by continuity of release rate at all phase transitions

where $0 < \rho < 1$. Modify the treatment of this section accordingly. Consider also the continuous time version of this situation.

(6.6) Note that continuity of u implies that $\dot{x} = 0$ at transition points.

6.4 GEOMETRIC PROGRAMMING

A simple problem that often turns up is that of minimizing the form $c_1 t + c_2 t^{-1}$ in $t \geqslant 0$, where c_1 and c_2 are positive coefficients. For example, if a sample of size t is being chosen, the cost of the sample will often be proportional to t, and the variance of an estimate derived from it proportional to t^{-1}, so that a cost function linear in these two components will be of the above form. Alternatively, a grab of capacity t will have to work for a time proportional to t^{-1} to excavate a given amount of material: both terms may enter in the cost function.

The solution is immediate, $t = (c_2/c_1)^{\frac{1}{2}}$, but an interesting fact, often remarked, is that this solution makes the two components of the cost function equal, each to $(c_1 c_2)^{\frac{1}{2}}$. This observation is the germ of the ingenious concept of *geometric programming* due to Zener (see Duffin, Peterson and Zener, 1967). Consider a cost function with m arguments and $m+1$ components, of the form

$$C(t) = \sum_{k=1}^{m+1} c_k \prod_{j=1}^{m} t_j^{a_{jk}}. \qquad (6.28)$$

Here the c_k are supposed positive, and the exponents a_{jk} of such value that $C(t)$ becomes infinite as any t_j approaches 0 or $+\infty$. Many practical cost functions can be represented in this form (which, with scant regard for etymology, Duffin *et al.* refer to as a *posynomial*); see Exercise (6.7). The above assumptions imply that $C(t)$ will reach its minimum in the finite interior of the region $t \geqslant 0$, and so at a stationary point. Let the kth component of cost be denoted

$$w_k = c_k \prod_j t_j^{a_{jk}} \quad (k = 1, 2, ..., m+1). \qquad (6.29)$$

Then the minimization equations can be written

$$\sum_k a_{jk} w_k = 0 \quad (j = 1, 2, ..., m). \qquad (6.30)$$

More convenient still is to work in terms of the proportion of the total cost due to the kth component,

$$x_k = \frac{w_k}{C} = \frac{w_k}{\sum w_k}, \qquad (6.31)$$

for which, in the optimal solution,

$$\sum x_k = 1,$$
$$\sum_k a_{jk} x_k = 0 \quad (j = 1, 2, ..., m). \qquad (6.32)$$

The identity, for x satisfying (6.32),

$$\prod_k x_k^{x_k} = \prod_k \left(\frac{c_k \prod_j t_j^{a_{jk}}}{C} \right)^{x_k} = \prod_k \left(\frac{c_k}{C} \right)^{x_k} \qquad (6.33)$$

then shows that the minimal total cost is equal to

$$\bar{C} = \prod_k (c_k/x_k)^{x_k}. \tag{6.34}$$

Thus the change of variable $t \to x$ has reduced the nonlinear problem of minimizing $C(t)$ to the linear problem of solving the system (6.32). Since the indices often have simple rational values, the system (6.32) is sometimes simple enough that it can be solved by inspection. With this once solved, the actual cost $\bar{C}$ and solution point $\bar{t}$ can be computed immediately from (6.34) and (6.29). Furthermore, these new variables x_k are meaningful, as being the proportions of the total cost due to the different cost-components in the optimal solution.

However, there is one drawback to this otherwise fortunate calculation. Only rarely will it happen in practice that the number of cost-components exceeds the number of independent variables by exactly one. However, by viewing the transformation $t \to x$ as effectively a passage to a dual problem, one can lighten the restriction.

Theorem (6.2)

Consider the maximization of

$$f(x) = \sum_{k=1}^{n} x_k \log(c_k/x_k) \tag{6.35}$$

in R_+^n subject to conditions (6.32). Suppose the c_k positive and the a_{jk} such that the system (6.32) is of full rank ($\leq n$) and possesses solutions in the interior of R_+^n. Then a modified form of the dual problem is: minimize

$$C(t) = \sum_{k=1}^{n} c_k \prod_{j=1}^{m} t_j^{a_{jk}} \tag{6.36}$$

in R_+^m. If the solutions and extrema reached in these respective problems are denoted $\bar{x}, \bar{t}$ and $\bar{f}, \bar{C}$ then

$$\bar{x}_k = c_k \prod_j \bar{t}_j^{a_{jk}}/\bar{C}, \tag{6.37}$$

$$\bar{C} = e^{\bar{f}} = \prod_1^n (\bar{x}_k/c_k)^{\bar{x}_k}. \tag{6.38}$$

So, a solution of the first problem provides a solution of the second, and this will usually be the easiest way of solving the second if $n-m$ is small relative to m. If $n = m+1$ then the first problem has but a single solution. If $n > m+1$ then one has a non-degenerate extremal problem, but with considerably fewer variables than the C-minimization problem.

Since f is concave, $\mathcal{X}\ (=R_+{}^n)$ convex, and the constraints linear, strong Lagrangian methods can certainly be used. Furthermore, it can be asserted that the multipliers y will be finite (i.e. $w = 1$, in the notation of Theorem (3.9)) and that the solution point is interior to $\mathcal{X}$ (i.e. $0 \ll \bar{x} \ll +\infty$). For, because of the assumption that the system (6.32) has full rank, bounded by n, and possesses solutions interior to $\mathcal{X}$, x can be perturbed about a solution point in such a way as to vary the b for this problem through a non-degenerate neighbourhood of its prescribed value; $b = (1, 0, 0, ..., 0)$. That is, this value is interior to $\mathcal{B}$. Further, finiteness of $\sum x_k$ implies finiteness of x, which implies finiteness of $f(x)$ and $U(b)$. These conditions ensure the possibility of normalization to $w = 1$.

Infinite values of x are excluded by finiteness of $\sum x_k$, as has just been remarked, and zero values are excluded by the fact that $-x_k \log x_k$ has infinite positive derivative at $x_k = 0$. If x_k were zero then a small perturbation into the interior of $\mathcal{X}$ consistent with (6.32) (possible, by hypothesis) would increase f.

Let Lagrangian multipliers $y_j = -\log t_j$ then be introduced to take account of the constraints (6.32). Since the solution $\bar{x}$ lies in the interior of $R_+{}^n$, the Lagrangian form

$$L = \sum x_k \log (c_k/x_k) - y_0(\sum x_k - 1) - \sum_{j=1}^{m} y_j \sum_k a_{jk} x_k \qquad (6.39)$$

has its maximum at the unique stationary point

$$\bar{x}_k = c_k t_0 e^{-1} \prod_j t_j{}^{a_{jk}}, \qquad (6.40)$$

and this maximum value is

$$\bar{L} = \sum \bar{x}_k + y_0 = t_0 e^{-1}\left(\sum_k c_k \prod_j t_j{}^{a_{jk}}\right) - \log t_0. \qquad (6.41)$$

The problem dual to the constrained maximization of f is the minimization of $\bar{L}$ with respect to t, and the maximum and minimum respectively attained will be equal. Now, carrying out the t_0 minimization, one finds that

$$t_0 = e/C(t), \qquad (6.42)$$

so that

$$\min_{t_0} \bar{L} = \log C(t). \qquad (6.43)$$

Substitution of (6.42) into (6.40) yields (6.37), and the fact that f equals the minimum of expression (6.43) with respect to t establishes equation (6.38).

Quite often a posynomial cost function $C(t)$ must be minimized subject to constraints on similar functions. The treatment of this section can be extended to such a case: see Exercise (6.8).

Expression (6.35) resembles the entropy of a distribution $\{x_k\}$, already considered in Exercise (2.5). In fact, the primal and dual versions of the geometric programming problem are immediately related to a physical-statistical problem: the determination of the relative abundances of chemical compounds in equilibrium. This is an application that will recur in Section (8.7), where we generalize the notion of a Lagrangian multiplier by regarding constrained extremal problems as limit cases of conditioned distributional problems.

Exercises

(6.7) The ore from a mine of known total extent is to be extracted by using a train of t_1 trucks, each of capacity t_2. The cost of a truck is proportional to t_2^α and the cost of haulage per year for such a train is $c_2 t_1^{\beta_1} t_2^{\beta_2}$, where the exponents are positive. If there is also a component of cost proportional to the number of trips that must be made, then the total cost is of the form

$$c_1 t_1 t_2^\alpha + c_2 t_1^{\beta_1} t_2^{\beta_2} + c_3 t_1^{-1} t_2^{-1}.$$

Determine the optimal values of t_1 and t_2. Consider also the effect of an additional term $c_4 t_2^\delta$, reflecting the cost of tunnel and track that will take trucks of capacity t_2.

(6.8) Suppose that in the theorem expression (6.35) is replaced by

$$f(x) = \sum_{k=1}^{n} x_k \log(c_k/x_k) + \sum_j S_j \log S_j,$$

where

$$S_j = \sum_{k \in A_j} x_k,$$

and the A_j are disjoint subsets of $(1, 2, ..., n)$. Show that the modified dual of this problem is of the form: minimize a posynomial subject to inequality constraints on other posynomials.

6.5 COMPLEMENTARY VARIATIONAL PRINCIPLES OF PHYSICS

There are many cases where either the equilibrium state or the dynamical evolution of a physical system can be expressed by a *variational principle*; by the requirement that the relevant variables adopt values which render extreme (at least locally) some function of these variables. For example, in the current flow example of Exercise (1.11), the flow takes the value minimizing the rate of dissipation of energy, consistent with applied potentials. An elastic body under strain will take a form which minimizes its total strain energy, consistent with applied deformations.

It is a temptation to speculate upon the basis of such natural "optimization" principles, but a temptation that lack of space bids us resist. However, they are certainly useful: not only do they give a concise and often co-ordinate-free characterization of the state (i.e. of the values of relevant variables), but they give an efficient means of approximately calculating the state: the variational or Rayleigh–Ritz method. For example, for the case of elastic deformation, a plausible deformation field of known functional form might be chosen, consistent with the constraints, but undetermined to within a few parameters. These parameters are then determined by requiring that a field of this form minimize the strain energy. In this way, excellent approximations to both field and the strain energy itself can be obtained.

There are cases where pairs of such principles exist, complementary in that the same quantity is characterized in one principle as the minimal value of some function, in the other as the maximal value of some other function. In such cases, application of the Rayleigh–Ritz method provides an approximation to the situation from two sides, and the closeness of the two bounds obtained in the two calculations gives a measure of the degree of accuracy of the approximations.

Now, the pattern of such complementary variational principles is very reminiscent of the relation between primal and dual problems and it is natural to ask whether this is a case of duality in the technical sense of Section (3.7). The relationship of duality is quickly seen in the particular case now to be described.

Consider a body being towed steadily through an incompressible viscous fluid, the fluid being stationary at infinity. Let $v_j(x)$ be the component of fluid velocity at x in the direction of x_j, the jth Cartesian component of x. Then v is prescribed on the surface of the body, and vanishes at infinity. The drag on the body is equal to

$$D = \frac{2\mu}{u} \int_E \sum \sum \varepsilon_{ij}^2 \, dx, \tag{6.44}$$

where μ is the coefficient of viscosity of the fluid, E is the space exterior to the body, ε_{ij} is a component of the rate-of-strain tensor,

$$\varepsilon_{ij} = \frac{1}{2}\left(\frac{\partial v_i}{\partial x_j} + \frac{\partial v_j}{\partial x_i}\right), \tag{6.45}$$

and u is the speed at which the body is being towed. Since the fluid is incompressible, the velocity field is divergenceless, which can be expressed by saying that

$$\sum_i \varepsilon_{ii} = 0. \tag{6.46}$$

Now, the first variational principle is that the velocity field v is such that the drag expressed by integral (6.44) is *minimal*, subject to (6.46) and prescription of v at the body surface.

An alternative expression of the drag is

$$D = -\frac{1}{2\mu u} \int_E \sum \sum \sigma_{ij}{}^2 \, dx + \frac{1}{u} \int_S \sum \sum \pi_j v_i \, dS_j, \tag{6.47}$$

where π_{ij} is a component of the stress tensor, σ_{ij} its traceless part

$$\sigma_{ij} = \pi_{ij} - \tfrac{1}{3}\delta_{ij} \sum_k \pi_{kk}, \tag{6.48}$$

S is the surface of the body, and dS_j the component of an element of surface normal to the x_j axis. The components of π satisfy

$$\sum_i \frac{\partial \pi_{ij}}{\partial x_i} = 0 \quad (j = 1, 2, 3) \tag{6.49}$$

The complementary variational principle is that π is such as to *maximize* expression (6.47), subject to condition (6.49). The physically relevant quantity, drag on the body, thus has the role of the common extremum of a maximum and a minimum problem. It is desired to show that one problem can be exhibited as the dual of the other.

Regard the primal problem, minimization of (6.44), as a problem in the two sets of variables v and ε, so that (6.45) now adopts the role of a constraint rather than of a definition. Associating multipliers $\pi_{ij}(x)$ and $\lambda(x)$ with constraints (6.45) and (6.46) we then have a Lagrangian form

$$L(v, \varepsilon; \lambda, \pi)$$

$$= \frac{2}{u} \int_E \left[\mu \sum \sum \varepsilon_{ij}{}^2 - \lambda \sum \varepsilon_{ii} + \tfrac{1}{2} \sum \sum \pi_{ij} \left(\frac{\partial v_i}{\partial x_j} + \frac{\partial v_j}{\partial x_i} \right) - \sum \sum \pi_{ij} \varepsilon_{ij} \right] dx$$

$$= \frac{2}{u} \int_E \left[\mu \sum \sum \varepsilon_{ij}{}^2 - \lambda \sum \varepsilon_{ii} - \sum \sum \pi_{ij} \varepsilon_{ij} - \sum \sum v_j \frac{\partial \pi_{ij}}{\partial x_i} \right] dx$$

$$+ \frac{2}{u} \int_S \sum \sum \pi_{ij} v_i \, dS_j. \tag{6.50}$$

Maximizing the form with respect to v, ε one sees that π must satisfy (6.49) and that

$$\pi_{ij} = 2\mu\varepsilon_{ij} + \lambda\delta_{ij}. \tag{6.51}$$

From condition (6.46) it follows that

$$\lambda = \tfrac{1}{3} \sum \pi_{kk} \tag{6.52}$$

so that (6.51) can be written

$$\sigma_{ij} = 2\mu \varepsilon_{ij}, \tag{6.53}$$

where σ is defined in terms of π by (6.48). The maximized value of the Lagrangian form is seen from (6.50)–(6.53) to be just expression (6.47), and the problem dual to the original one would certainly be that this expression should be maximal, subject to (6.48) and (6.49). Relation (6.53) enables us to identify π, hitherto merely a formal multiplier, as the stress tensor.

Of course, primal and dual will not have a common extremum unless Lagrangian methods are applicable. However, the positive definite quadratic nature of form (6.44) plus the linearity of constraints (6.45) and (6.46) guarantees validity.

An attractive possibility is that a natural variational principle could be coupled with an imposed optimality condition. For example, in the case just considered, v must be such as to minimize drag, and one might well wish to choose the body form itself so as to minimize drag. One is thus minimizing expression (6.44) with respect to both velocity field and body form; a compound variational formulation amenable to Rayleigh–Ritz methods. This approach is being pursued by Dr. S. R. Watson of Cambridge University. Dr. Watson also brought this particular example of a pair of complementary principles to my attention, for which I am grateful.

A more general treatment of complementary variational principles seems to belong with a discussion of the min–max theorem: see Section (10.4).

CHAPTER 7

Some Problems with Linear Constraints. II

7.1 SOME SIMPLE CONTROL PROBLEMS

The dynamic optimization problems of Sections (2.5), (5.1) and (6.3) all constitute examples of optimal control: the control of a deterministic process over a prescribed time interval in discrete time. As we have said on each occasion, dynamic problems do lie outside the scope of this volume. Nevertheless, it is appropriate to demonstrate one main point: the relationship of the strong Lagrangian principle to the *Pontryagin maximum principle* of control theory.

The desirable canonical form for a control problem of this type is, that there is a *state variable* x whose value x_{t+1} at time $t+1$ is determined by a simple recursion

$$x_{t+1} = r_t(x_t, u_t). \tag{7.1}$$

Here u_t is the value of the *decision* or *control* variable at time t, r is a function of the variables t, x_t and u_t, and the sequence $\{u_t\}$ is to be chosen so as to maximize the value of some criterion function

$$f(x, u) = f(x_0, x_1, ..., x_N; u_0, u_1, ..., u_{N-1}).$$

This maximization is subject to the sequence of restrictions (7.1), to any other overriding limitation on the domain of x and u, and, usually, to the prescription of the initial value x_0. The state and control variables, x_t and u_t, will usually be vectors, of different dimensionalities.

So, for the dam regulation problem of Section (6.3), the recursion (7.1) is just the balance relation (6.15), x and u being respectively amount of water held and the amount released. For the cross-current extraction problem of Section (2.5), the recursion is given in implicit form by (2.29). The economic programming problem of Section (5.1) obviously has a simply recursive character, although some reformulation is needed to bring it into the canonical form (7.1); see Section (8.2).

Suppose the restrictions of the problem, apart from recursion (7.1), confine the joint variable (x, u) to a region $\mathcal{W}$. If the situation is such that the sequence of constraints implied by the recursion (7.1) can be taken account of by strong Lagrangian methods, then there exists a sequence $\{y_t\}$ such that (x, u) maximizes the Lagrangian form

$$L(x, u; y) = f(x, u) - \sum_{t=0}^{N-1} y_t^{\mathrm{T}}(x_{t+1} - r_t(x_t, u_t)) \qquad (7.2)$$

in $\mathcal{W}$.

Provided appropriate derivatives exist, y_t^{T} will have the interpretation $\partial U/\partial b_t$, where b_t is the value of the bracket y_t^{T} multiplies in (7.2). That is, y_t^{T} can also be regarded as $\partial U/\partial x_{t+1}$: the rate of change of maximal utility achievable with change in the value of the state variable at time $t+1$, if this value could be varied by an externally applied perturbation at this point. In other words we regard (7.1) as modified to

$$x_{t+1} = r_t(x_t, u_t) + b_t, \qquad (7.3)$$

where the "forcing" term b_t can then directly influence x_{t+1}. For example, for the dam problem, if an additional source could supply an amount b_t of water between instants t and $t+1$ then (6.15) would become modified to

$$x_{t+1} = x_t + v_t - u_t + b_t,$$

and y_t would be the "fair price" for such additional water.

In dynamic problems f will often have the form

$$f(x, u) = \sum_0^{N-1} h_t(x_t, u_t) + k(x_N). \qquad (7.4)$$

That is, the utility is made up additively of contributions from each instant of time, the contribution depending only upon current values of state variable, control variable and time. Such a form often occurs naturally in control problems, or can be achieved by an appropriate choice of variables.

In such a case, then, the optimal value of u_t will minimize the expression

$$L_t = L_t(x_t, u_t; y_t) = h_t(x_t, d_t) + y_t^{\mathrm{T}} r_t(x_t, u_t), \qquad (7.5)$$

in which the two individual terms can be regarded as the *current* and *future* components of utility derived from a decision u_t, respectively. Furthermore, the value of x_t must maximize the expression

$$L_t(x_t, u_t; y_t) - y_{t-1}^{\mathrm{T}} x_t \quad (t = 1, 2, ..., N).$$

If this maximum is attained at a stationary point then

$$y_{t-1} = \frac{\partial}{\partial x_t} L_t(x_t, u_t; y_t), \qquad (7.6)$$

a relation which provides a recursion for the $\{y_t\}$ sequence. Note that the initial recursion (7.1) can formally be written

$$x_{t+1}{}^{\mathrm{T}} = \frac{\partial}{\partial y_t} L_t(x_t, u_t; y_t),\qquad(7.7)$$

which gives it an interestingly conjugate relationship to (7.6).

The statement that u_t is to be chosen so as to maximize expression (7.5), where the sequence $\{y_t\}$ is, with exceptions, generated by (7.6), constitutes the *Pontryagin maximum principle* in discrete time. The point of the principle is that it provides a reduction of the optimization problem: one chooses the value of u_t individually so as to maximize L_t, without explicitly worrying about the values assigned to u at other time points. Of course, the values of u chosen at different points of time must be inter-related, but the inter-relation is provided by the occurrence of x_t and y_t in L_t, themselves recursively related. So, the task of optimizing with respect to a whole sequence is replaced by a sequence of "instantaneous" optimizations.

Determination of the solution by maximization of a Lagrangian form (7.2), and hence the Pontryagin principle, will be valid if (x, u) may vary in a convex set $\mathscr{W}$, the criterion function f is concave in this set, the relations (7.1) are linear, and some regularity condition is satisfied which ensures that the y_t will be finite.

The most irksome of these conditions is the requirement that relations (7.1) be linear. For example, this will virtually never be satisfied for the cross-current extraction problem of Section (2.5). However, it seems that this restriction cannot be greatly relaxed, and the discrete principle will often be valid only in the weak sense, that $L(x, u; y)$ is stationary at the solution point $(\bar{x}, \bar{u})$. In the case of continuous time the essential validity of the strong maximum principle can be seen by direct arguments, not based at all on notions of convexity, etc. This makes one inclined to believe that the maximum principle is essentially a continuous-time result, with a complete discrete-time analogue only in rather special cases.

The treatment of the dam problem in Section (6.3) amounted exactly to use of the discrete maximum principle. The extreme values $x = 0$, C of the state variable were usually attained, so that in this sense the solution generally lay on the boundary of $\mathscr{W}$. Consider now a problem in which it is restrictions on the values of the control variable rather than of the state variable which are important.

We consider a very simple model of trajectory optimization for a rocket, in discrete time. Let x describe the dynamical co-ordinates of the rocket (position, velocity, etc.) and u the applied vector thrust, and suppose that the dynamical equations of the rocket (in discrete time) then have the linear form

$$x_{t+1} = Ax_t + Bu_t.\qquad(7.8)$$

The linearity assumption implies, for example, that mass-variation is neglected, and also nonlinear variation of gravitational field with position. These rather unrealistic limitations can be removed in a fuller, continuous-time treatment, but the present simple model still brings out a number of the essential features of problem and solution.

We shall suppose that one wishes to maximize some linear function, $c^T x_N$, of the terminal co-ordinate x_N, subject to (7.8), prescription of x_0, and the inequalities

$$|u_t| \leqslant M_t \quad (t = 0, 1, ..., N-1), \tag{7.9}$$

$$\sum_0^{N-1} |u_t| \leqslant K, \tag{7.10}$$

where the values M_t, K are given. The quantity $c^T x_N$ might represent, for example, the height of the rocket above ground at time N. Restriction (7.9) (which, together with prescription of x_0, will be regarded as specifying $\mathscr{W}$) states that thrust has a time-variable upper bound (reflecting physical limitations of the fuel supply and combustion mechanisms). Inequality (7.10) could be regarded as a bound on the total amount of fuel consumed. This constraint will be accounted for by introduction of a positive Lagrangian multiplier λ (justifiable, since the function $\sum |u_t|$ is convex: see Theorem (3.10)).

The problem of reaching maximal height at time N, with limitations on total fuel consumption and rate of fuel consumption, can thus be formulated as the maximization of

$$f(x, u) = c^T x_N - \lambda \sum_0^{N-1} |u_t| \tag{7.11}$$

subject to constraints (7.8), within the region $\mathscr{W}$ specified by (7.9) and prescription of x_0.

Thus, the optimal thrust u_t at time t maximizes

$$L_t = -\lambda |u_t| + y_t^T (A x_t + B u_t) \quad (y_t = 0, 1, ..., N-1) \tag{7.12}$$

and x_t maximizes $L_t - y_{t-1}^T x_t$ ($t = 1, 2, ..., N-1$), or $(c - y_{N-1})^T x_N$ for $t = N$. Since the values of x are unrestricted, but necessarily finite, these x-maxima must be reached at stationary points, so that

$$y_{t-1} = A^T y_t \quad (t < N), \tag{7.13}$$

$$y_{N-1} = c \tag{7.14}$$

The y_t sequence is thus completely determined,

$$y_t = (A^T)^{N-t-1} c, \tag{7.15}$$

and one sees that u_t must maximize $\zeta_t^T u_t - \lambda |u_t|$ where

$$\zeta_t^T = c^T A^{N-t-1} B. \tag{7.16}$$

This expression is maximized for a given value of $|u_t|$ if u_t is chosen in the same direction as ζ_t, i.e.

$$u_t = \frac{|u_t|}{|\zeta_t|} \zeta_t. \tag{7.17}$$

The expression to be maximized then reduces to $|u_t|(|\zeta_t| - \lambda)$, and $|u_t|$ will be plainly taken equal to its maximal or minimal value according as the bracket is positive or negative. That is, the optimal thrust value is

$$u_t = \left\{ \begin{array}{c} \dfrac{M_t \zeta_t}{|\zeta_t|} \\ \text{indeterminate} \\ 0 \end{array} \right\} \quad \text{according as} \quad |\zeta_t| \left\{ \begin{array}{c} > \\ = \\ < \end{array} \right\} \lambda, \tag{7.18}$$

where ζ_t is given by (7.16).

This conclusion demonstrates the rather extreme form of control, "bang–bang control", which is optimal if u_t enters the criterion function by a term proportional to $|u_t|$. With minor exceptions (see Exercise (7.1)) either maximal thrust is exerted in the current direction of ζ or no thrust is exerted at all.

The multiplier λ either takes the non-negative value which assures equality in (7.10), or, if this is not possible, takes the value zero. A zero value of λ is an indication that additional fuel would not improve performance—usually because it is physically impossible to use all fuel available in the given time (i.e. $\sum M_t \leqslant K$).

The three cases distinguished in (7.18) are known as régimes of maximal thrust, intermediate thrust and null-thrust (or coasting) respectively. As is noted in Exercise (7.1), the case of intermediate thrust occurs only rather exceptionally in this linear model, although it may be more frequent for non-linear models. For example, if the rocket suffers a drag which increases rapidly with velocity at low altitudes, where the atmosphere is dense, then a positive but not maximal thrust may be desirable at these altitudes.

Exercises

(7.1) Note that, if the $n \times n$ matrix A has distinct eigenvalues θ_i, then one can write

$$|\zeta_t|^2 = \sum \sum \gamma_{ij} (\theta_i \theta_j)^{N-t}$$

where ζ_t is defined in (7.16). Show that $|\zeta_t|^2 = \lambda$ holds for all t if it holds for more than n^2 values of t. That is, conditions for intermediate thrust hold always if they hold for more than n^2 instants of time.

(7.2) Work through the case

$$A = \begin{pmatrix} 1 & 1 \\ . & 1-\alpha \end{pmatrix}, \quad B = \begin{pmatrix} 0 \\ 1 \end{pmatrix}, \quad c = \begin{pmatrix} 1 \\ 0 \end{pmatrix}$$

in detail, where α lies in $(0, 1)$. Here the first component of x corresponds to vertical height, the second to vertical velocity, and there is a drag term $\alpha \times$ velocity.

7.2 THE OPTIMIZATION OF BRACED FRAMEWORKS; MICHELL STRUCTURES

The design of the most economical framework which will carry a given load provides an example of great intrinsic interest, which also demonstrates the power of Lagrangian methods.

Suppose there are a number of points in physical space, with vector co-ordinates x_i $(i = 1, 2, ..., n)$ which can serve as the hinged joints of a framework. That is, these points, which will be termed *nodes*, can be connected by members whose ends are freely hinged, and so are in pure tension (ties) or compression (struts or braces).

It will be assumed that some of the nodes are fixed in space, in that they will not move under any load likely to be placed on them. These are the *foundation points*. The potential foundation points often form a surface which will be referred to as the *foundation arc*. A node which is not a foundation point will be termed simply a *joint*.

An external prescribed load is applied to some or all of the nodes: the function of the framework is to communicate this load to the foundation. The problem is, to find the most economic framework within a prescribed region of space which will do this, without failure or excessive deformation.

As examples, a bridge is a framework which communicates a load distributed along its deck to a foundation consisting of available parts of the river banks and bed; see Figures (7.1) and (7.2). A coat-hook could be conceived as a framework carrying a vertical point load, see Figure (7.3), which it communicates to the foundation consisting of the wall. One will generally require, for aesthetic, practical and constructional reasons, that the coat-hook be relatively compact in form. For example, one would wish to exclude a web-like structure which permeated the universe, although such a structure might be very economic of material. This is the reason for the possible restriction to a given region of space, mentioned above.

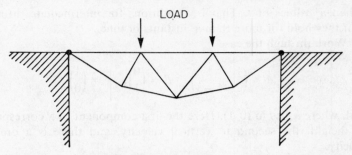

Figure 7.1 A bridge (truss) viewed as a framework communicating the load applied to its deck to foundation points on the river banks

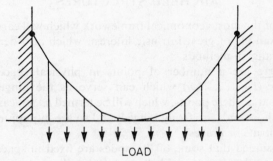

Figure 7.2 A bridge (suspension) viewed as a framework communicating the load applied to its deck to foundation points on the gorge walls

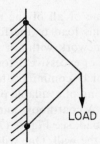

Figure 7.3 A schematic coathook, viewed as a framework communicating a vertical point-load to the foundation arc constituted by a vertical wall

The measures open to us in our optimization are:
(a) To vary the cross-sections of members connecting the nodes.
(b) To vary the position and number of nodes within the permitted space (load- and foundation-points remaining fixed).

(c) To vary the materials of the framework, and

(d) To vary the pre-stressing of the framework.

Regarding (b), at most two materials will be considered; one for tension members and one for compression members. To avail oneself of all the freedom offered by the other variations in fact makes optimization easier and simpler: it is characteristic of extremal problems that in general they become simpler as one enlarges the field of possible variations. Variations (a) and (d) will be considered first, and then (b). That is, we begin with a framework of fixed geometry, which is later modified.

Let F_i be the external load applied at node i: a vector force. Let P_{ij} be the signed scalar force in member ij (the member joining nodes i and j): positive for a tension and negative for a compression. Then the equilibrium equation at point i is

$$\sum_j P_{ij} u_{ij} = F_i, \tag{7.19}$$

where u_{ij} is the unit vector in the direction of $x_i - x_j$. Relation (7.19) holds only at joints, and not at foundation points, where there will be an equilibrating reaction from the foundation.

If the framework is statistically determinate, then relations (7.19) determine the P_{ij}. If indeterminate, then the P_{ij} can only be determined if one also considers the deformations in the framework caused by the load, and brings in the concepts of elasticity theory, relating stresses and strains (forces and deformations). However, this complication is completely avoided if, in accordance with (d), the idea of an arbitrary pre-stressing is allowed. For, if the structure is not required to be stressed (or unstressed) in any particular way in the unloaded state, then any set of forces P_{ij} compatible with (7.19) is permissible. That is, we can regard the P_{ij} as a set of independent variables, subject only to constraints (7.19).

Suppose K_T and K_C are the limiting magnitudes of tension and compression per unit cross-section that can be sustained with the materials available, so that

$$-K_C a_{ij} \leqslant P_{ij} \leqslant K_T a_{ij}, \tag{7.20}$$

where a_{ij} is the cross-section of member ij. (Actually, the yield point may not depend solely on forces per unit cross-section: this point is considered in a moment.) It is plain that, if yield is the only consideration (rather than excessive deformation), then for economy one will design to the limit, and one of the bounds in (7.20) will be attained for each member. Thus, the volume of the framework which is in tension, for example, will be

$$\sum_T a_{ij} r_{ij} = K_T^{-1} \sum_T r_{ij} P_{ij}, \tag{7.21}$$

where r_{ij} is the length of member ij,

$$r_{ij} = |x_i - x_j|, \tag{7.22}$$

and $\sum_T$ covers all members in tension. If the costs per unit volume of the tension and compression materials are L_T and L_C respectively, then the total cost of the material of framework members is

$$f(P) = \alpha_T \sum_T r_{ij} P_{ij} - \alpha_C \sum_C r_{ij} P_{ij}, \tag{7.23}$$

where $\alpha_T = L_T/K_T$, etc., and $\sum_C$ covers members in compression. It is quantity (7.23) that will be minimized, as a function of the P_{ij}, subject to restrictions (7.19).

The advantages of envisaging an arbitrary prestressing are now apparent. The P_{ij} are then subject only to the constraints (7.19); it is then natural to take the P_{ij} as independent variables rather than the a_{ij}, which might otherwise have been taken. The a_{ij} are in fact eliminated at an early stage in the calculation, by the requirement that each member be stressed to whichever of the limits (7.20) is appropriate.

There are several simplifications in our treatment which might not always be acceptable. The self-weight of the structure has been neglected, and also the costs of joints and of fabrication generally. To include these factors would not be impossible, but they can reasonably be left out of a first treatment. There is another point, which could be serious. To assume that it is the magnitude of force per unit cross-section which dictates failure might be reasonable for tension members, but not necessarily for members in compression. A strut can fail by buckling, and the longer the strut (for a given cross-section and compressive force) the more likely it is to do so. Some allowance for this feature should probably be made in the cost function. However, it will be seen that the solution to the problem as posed tends in many cases to avoid long members, and so at least not produce a structure excessively prone to buckling.

The problem to be solved, maximization of (7.23) subject to (7.19), is almost a linear one, but not quite. The coefficient of P_{ij} in $f(P)$ depends upon the sign of P_{ij}. To this extent, $f(P)$ is nonlinear. However, it is at least convex (as is appropriate for a minimization problem), and we can apply Lagrangian methods. Let Lagrangian multipliers λ_i be introduced to take account of the constraints (7.19). Since these are vector constraints, the λ_i will also be vector. Indeed, if $U(F)$ is the minimal cost of a framework carrying loads F, then

$$\lambda_i^k = \partial U/\partial F_i^k, \tag{7.24}$$

wherever the derivative exists, λ_i^k and F_i^k being the kth components of their respective vectors. That is, λ_i is the rate of change of cost with load at node i

(in the vector sense). The value of λ_i will be zero at a foundation-point: formally, because (7.19) does not hold at a foundation-point; intuitively, because a foundation-point can take any load without cost.

The Lagrangian form for the problem is

$$L(P, \lambda) = \alpha_T \sum_T r_{ij} P_{ij} - \alpha_C \sum_C r_{ij} P_{ij} + \sum_i \lambda_i \cdot \left(F_i - \sum_j P_{ij} u_{ij} \right). \tag{7.25}$$

The physical convention has been employed, of denoting an inner product $\sum \lambda^k F^k$ by $\lambda . F$ rather than $\lambda^T F$.

Theorem (7.1)

The problem dual to the design problem is: choose the λ_i to maximize

$$\Phi(\lambda) = \sum_i \lambda_i \cdot F_i, \tag{7.26}$$

subject to

$$\lambda_i = 0 \quad \text{on the foundation,} \tag{7.27}$$

and

$$-\alpha_C r_{ij} \leqslant (\lambda_i - \lambda_j) . u_{ij} \leqslant \alpha_T r_{ij} \quad (i, j = 1, 2, ..., n). \tag{7.28}$$

A tension (compression) member may exist between nodes i and j only if the right-hand (left-hand) bound is attained in (7.28).

The virtue of considering the dual is that variation (b) (variation of the framework geometry) is much more easily accomplished in the dual than in the primal, as will be seen. The interpretation of the dual is: the costs λ_i shall be chosen to maximize the cost (7.26) of accepting the prescribed load-pattern F_i, subject to (7.27), and to the requirement (7.28) that an additional tension δ in member ij shall not produce an additional cost exceeding $\alpha_T r_{ij} \delta$; correspondingly for compressions.

The theorem follows immediately from minimization of L with respect to P. Minimizing with respect to P_{ij}, we find that the right-hand equality must hold in (7.28), or the left-hand equality, or both inequalities, according as P_{ij} is positive, negative or zero. The minimized value of L left is just (7.26), and it is known from Section (3.7) that this must be maximal with respect to the λ_i.

Now, suppose that the possible node-positions are allowed to become dense in the permitted region S of x-space. That is, nodes may exist anywhere in S, and any pair of nodes may conceivably be connected by a member. In this way variation (b) is achieved. One is effectively allowing the possibility of a "continuous framework"; a lattice or cellular structure of variable properties and indefinitely fine scale.

The quantity λ_i will now be defined, not merely at the discrete points x_i, but for all x in S, and will constitute a vector field $\lambda(x)$ in this region. The quotient effectively occurring in (7.28) can be written

$$\frac{(\lambda_i - \lambda_j) \cdot u_{ij}}{r_{ij}} = \frac{\Delta\lambda \cdot \Delta x}{|\Delta x|^2}, \tag{7.29}$$

where $\Delta x = x_i - x_j$ is a vector increment in position, and $\Delta\lambda = \lambda_i - \lambda_j$ the corresponding increment in λ. The quantity of interest is the limiting value of this ratio:

$$\frac{D\lambda}{Dx} = \lim_{|\Delta x| \to 0} \frac{\Delta\lambda \cdot \Delta x}{|\Delta x|^2}, \tag{7.30}$$

the rate of change in a certain x direction (the direction of Δx) of the component of $\lambda(x)$ in that direction. Theorem (7.1) then has an obvious limit form:

Theorem (7.2)

The problem dual to the continuous design problem is: choose $\lambda(x)$ to maximize

$$\Phi(\lambda) = \sum_i \lambda(x_i) \cdot F_i \tag{7.31}$$

subject to $\lambda(x) = 0$ on the foundation, and

$$-\alpha_C \leqslant \frac{D\lambda}{Dx} \leqslant \alpha_T \quad (x \in S). \tag{7.32}$$

Tension (compression) members may exist only on line elements for which the right-hand (left-hand) bound is attained in (7.32).

Expression (7.31) will be replaced by an integral if loads are distributed.

This is very close to the theorem obtained by Michell (1904) in a brilliant and very early paper. Michell necessarily used different methods, and an argument which was not quite complete, although followed through to detailed design conclusions in a masterly fashion. The paper would have been impressive had it appeared fifty years later, and must rank as one of the foremost and most original contributions to the optimization literature. It also foreshadowed the modern theory of plastic solids, as we shall see.

Michell does not have the dual characterization of the field $\lambda(x)$ (that it maximizes expression (7.31)), and the nature of $\lambda(x)$ itself is somewhat unclear in his treatment. He refers to it as a "virtual deformation": one of the more puzzling concepts of classical statics. All that $\lambda(x)$ has in common with a physical deformation is that both are vector fields—of which more anon. The correct interpretation of $\lambda(x)$ is the "price" characterization, (7.24). Since λ is the multiplier associated with a vector balance condition, it also has the character of a vector potential: the value of P_{ij} maximizing the Lagrangian form depends on $\lambda_i - \lambda_j$.

Theorem (7.2) has immediate implications for the form of the structure: the conclusions of the next two theorems are also due to Michell.

Theorem (7.3)

The differential coefficient $D\lambda/Dx$ achieves its distinct extrema in directions which are mutually orthogonal. Hence, if tension members meet compression members, they must do so at right-angles.

For, if in some Cartesian co-ordinate system x, λ and Δx have components x^i, λ^i and Δ^i then

$$\frac{\Delta x . \Delta \lambda}{|\Delta x|^2} = \frac{\sum_i \Delta^i \sum_j \left(\frac{\partial \lambda^i}{\partial x^j} \Delta^j + o(\Delta)\right)}{\sum_i (\Delta^i)^2}, \tag{7.33}$$

so that

$$\frac{D\lambda}{Dx} = \frac{u^T E u}{u^T u}, \tag{7.34}$$

where u is a vector in the direction of Δx, and E is the symmetric matrix with elements

$$e_{ij} = \frac{1}{2}\left(\frac{\partial \lambda^i}{\partial x^j} + \frac{\partial \lambda^j}{\partial x^i}\right). \tag{7.35}$$

The extreme values of the ratio (7.34) are just the eigenvalues of E, and the values of u for which these are achieved are the corresponding eigen-vectors. Eigen-vectors corresponding to distinct eigenvalues are certainly mutually orthogonal (see, for example, Bellman (1960), p. 36) so the assertions of the theorem follow.

It is seen from the theorem that it is impossible, for example, that three members of an optimal structure should form a triangle, unless they were all in tension, or all in compression. For structures which are wholly in tension (or compression) there is a very simple result, due to Clerk Maxwell (see Exercise (7.6)).

Theorem (7.2) has further implications, which will be investigated only for the case of two-dimensional structures.

Theorem (7.4)

The following are the possibilities at a point in a two-dimensional framework:

(a) *The eigenvalues of E lie strictly between $-\alpha_C$ and α_T. Then no member passes through the point.*

(b) *One eigenvalue only attains one of the bounds (7.32), say α_T $(-\alpha_C)$. Then any member through the point is in the direction of the corresponding eigen-vector and is under tension (compression). It thus either terminates at the point or passes through it in a straight line.*

6

(c) *The eigenvalues both equal the same extreme value, say α_T ($-\alpha_C$). Then members may extend from the point in any direction, but are all in tension (compression).*

(d) *The two eigenvalues equal α_T and $-\alpha_C$. Then tension and compression members may extend from the point in the directions of the corresponding eigen-vectors and are consequently orthogonal. If this condition holds in a region R, then tension and compression members form mutually orthogonal nets in R. These two nets have the additional property; that any tension (compression) line connecting two given compression (tension) lines turns through a constant angle in passing from one line to the other (see Figure (7.4)).*

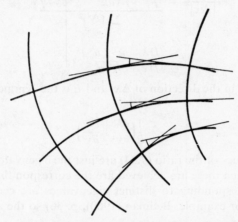

Figure 7.4 Curves of a Hencky–Prandtl net, illustrating the property that all curves of one family turn through a constant angle along the arc lying between two given curves of the other family

All statements except the last are immediate consequences of previous theorems. The last assertion sets a significant further limitation on the form of the structure.

Note that "members" are now infinitesimal line-elements, but that a chain of such elements forms a curve, these curves being the tension and compression lines referred to in the last part of the theorem. It is these two families of curves which constitute the two mutually orthogonal nets in R.

The proof of the final assertion requires further consideration of the properties of the vector field $\lambda(x)$. For illustrative purposes, it is helpful to think of $\lambda(x)$ as a definite physical quantity, such as the displacement of the point of a continuum with initial co-ordinate x. Of course, in the original problem it had no such interpretation—the "displacement" must be a conceptual one in a conceptual space. As mentioned before, $\lambda(x)$ is just what is referred to classically in this context as a "virtual deformation", which is highly misleading, since it does not correspond to any physical deformation of the original structure at all. However, it has the same transformation properties as a deformation field, under change of co-ordinates, so one gains in vividness—if at the risk of some confusion—by thinking of it in such terms.

Consider the matrix

$$\Lambda = (\lambda_{ij}) = \left(\frac{\partial \lambda^i}{\partial x^j}\right) \tag{7.36}$$

in the two-dimensional case. If the x-space is rotated by the operation $x \to Hx$ (so that H is orthogonal) then Λ suffers the transformation $H\Lambda H^T$. The invariants of such a transformation—which must be the physically significant local aspects of the λ-field—are just

$$w = \tfrac{1}{2}(\lambda_{12} - \lambda_{21}) \tag{7.37}$$

(the skew component of Λ) and the eigenvalues of $E = \tfrac{1}{2}(\Lambda + \Lambda^T)$.

The quantity w measures the *rotation* suffered by a line element in x-space when the λ-deformation is applied. The eigenvalues of E measure the *dilation* (i.e. contraction or extension) which is suffered, in the canonical directions in which this dilation is extremal. These are just the directions in which elements of the optimal framework may be placed.

Suppose that these canonical directions (necessarily orthogonal) make angles $\phi, \phi + \tfrac{1}{2}\pi$ with the x^1 axis, and correspond to dilations $\alpha_T, -\alpha_C$. If

$$c = \cos\phi, \quad s = \sin\phi, \tag{7.38}$$

then E must have the spectral representation

$$E = \alpha_T \binom{c}{s}\binom{c}{s}^T - \alpha_C \binom{-s}{c}\binom{-s}{c}^T, \tag{7.39}$$

so that

$$\left.\begin{aligned}
e_{11} &= \alpha_T \cos^2\phi - \alpha_C \sin^2\phi, \\
e_{22} &= \alpha_T \sin^2\phi - \alpha_C \cos^2\phi, \\
e_{12} &= e_{21} = (\alpha_T + \alpha_C)\cos\phi\sin\phi.
\end{aligned}\right\} \tag{7.40}$$

Consider now how ϕ varies as we move along a tension or compression member (i.e. move in a direction of maximal or minimal dilation under the deformation). By doing this we shall be moving along a member of one of two families of mutually orthogonal curves. It is convenient to transform to curvilinear co-ordinates $\xi(x), \eta(x)$ so that these two families of curves are described by $\xi = \text{const}, \eta = \text{const}$.

Suppose we are at a point at which the ξ and η directions coincide locally with the x^1 and x^2 directions respectively, so that $\phi = 0$—this can always be achieved by an initial rotation of the co-ordinate system. Then E takes the form

$$E = \begin{pmatrix} \alpha_T & \cdot \\ \cdot & -\alpha_C \end{pmatrix} \tag{7.41}$$

at the point. Suppose now that we move an infinitesimal distance δx^1 along the x^1-axis (i.e. the direction of the tension member) and that E then transforms to

$$E + \delta E = \begin{pmatrix} \alpha_T & \cdot \\ \cdot & -\alpha_C \end{pmatrix} + \delta x^1 \begin{pmatrix} \beta_{11} & \beta_{12} \\ \beta_{21} & \beta_{22} \end{pmatrix} + o(\delta x^1). \tag{7.42}$$

Thus $\beta_{ij} = \partial e_{ij}/\partial x^1$, so that

$$\begin{aligned} \beta_{11} &= \lambda_{111}, \\ \beta_{22} &= \lambda_{221}, \\ \beta_{12} &= \beta_{21} = \tfrac{1}{2}(\lambda_{121} + \lambda_{211}), \end{aligned} \tag{7.43}$$

where

$$\lambda_{ijk} = \frac{\partial^2 \lambda^i}{\partial x^j \, \partial x^k}. \tag{7.44}$$

Now, by the properties of the λ-field, $E + \delta E$ must have the same eigenvalues as E: namely α_T and $-\alpha_C$. So, in particular, E and $E + \delta E$ must have the same trace and determinant. These conditions imply that

$$\beta_{11} = \beta_{22} = 0, \tag{7.45}$$

or

$$\lambda_{111} = \lambda_{221} = 0, \tag{7.46}$$

and, by the same argument in the x^2 direction

$$\lambda_{112} = \lambda_{222} = 0. \tag{7.47}$$

The expression for e_{12} in (7.40) implies the identification

$$\beta_{12} = \frac{\partial e_{12}}{\partial x^1} = (\alpha_T + \alpha_C) \frac{\partial \phi}{\partial x^1}. \tag{7.48}$$

It follows then from (7.43), (7.46), (7.47) and (7.48) that

$$(\alpha_T + \alpha_C) \frac{\partial \phi}{\partial x^1} = \tfrac{1}{2}(\lambda_{121} + \lambda_{211}) = \tfrac{1}{2}(\lambda_{211} - \lambda_{121}) = -\frac{\partial w}{\partial x^1}. \tag{7.49}$$

Now, the only significance of the x^1-axis is that it coincides locally with the ξ-axis: the universal expression of (7.49) would thus be

$$\frac{\partial}{\partial \xi}(w + \sigma\phi) = 0, \tag{7.50}$$

where

$$\sigma = \alpha_T + \alpha_C. \tag{7.51}$$

By a similar argument along the x^2-axis it is seen that

$$\frac{\partial}{\partial \eta}(w - \sigma\phi) = 0. \tag{7.52}$$

That is, $w + \sigma\phi$ is a function of η alone, say $B(\eta)$, and $w - \sigma\phi$ a function of ξ alone, say $A(\xi)$. The angle ϕ, as a function of ξ, η, then has the representation

$$\phi(\xi, \eta) = \frac{1}{2\sigma}(B(\eta) - A(\xi)). \tag{7.53}$$

Thus, for a pair of values ξ and ξ', the difference $\phi(\xi, \eta) - \phi(\xi', \eta)$ is independent of η, which is the final assertion of Theorem (7.4).

Detailed applications of these results are developed in the next section.

Orthogonal nets having the property described in the last sentence of Theorem (7.4) are often referred to as *Hencky–Prandtl nets*. They seem to have been recognized first by Michell, in this context, in his pioneering paper of 1904. However, the shear-lines in perfectly plastic solids are also described by such nets (for very good reasons), and it was in this context that Hencky and Prandtl rediscovered them in 1923.

Exercises

(7.3) Consider how the treatment of this section must be modified, if the structure is required to carry any one of a number of alternative loading patterns (see Chan, 1968).

(7.4) Consider how the treatment should be modified if the self-weight of the structure is added to the load.

(7.5) Suppose that the loads on a structure are in overall static balance, so that the structure equilibrates these loads, without a foundation. Suppose also that each member is stressed to the limit. Then show from (7.19) and (7.20) that $\sum x_i . F_i = K_T$ (volume of material in tension) $- K_C$ (volume of material in compression).

(7.6) (Maxwell). Suppose that in the case of Exercise (7.5) a structure exists which is wholly in tension (or compression). Show from the result of Exercise (7.5) that this structure is optimal.

(7.7) Show that in the case of Exercise (7.6) the λ-field solving the dual problem is an appropriate pure "dilation": i.e. $\lambda(x)$ equals a constant plus $\alpha_T x$ or $-\alpha_C x$, according as the structure is wholly in tension or compression. Interpret.

(7.8) Consider the analogue of Theorem (7.4) in more than two dimensions.

(7.9) Show that, if one of the ξ-lines is straight, then so are they all.

(7.10) Show that the "rotation" w is the marginal cost of strengthening the structure to cope with a locally applied couple (turning force).

7.3 SOME EXAMPLES OF MICHELL STRUCTURES

An analytic solution for the λ-field characterized in Theorem (7.2), and hence for the structure, can be obtained in only a few cases. However, the more explicit properties of the λ-field determined in Theorems (7.3) and (7.4) give an intuitive lead in the building up of almost or fully optimal solutions.

Consider a part of a plane framework which is not directly (i.e. externally) loaded, and which contains both tension and compression members. By Theorems (7.3) and (7.4) these two sets of members then constitute a Hencky–Prandtl net. Examples of such nets are the rectangular nets and circular fan-shaped nets of Figure (7.5) and the families of conjugate equi-angular spirals of Figure (7.8). In this last case the spirals have equations

$$r = ae^{c\theta},$$

$$r = be^{-\theta/c}$$

in polar co-ordinates (r, θ) based on their common origin. The constant c is common, and the two families of spirals are parametrized by a and b. The optimal structure can sometimes be built up by piecing together sections of such nets. For example, the net around a load point may well be locally of the circular fan form, with origin at the load point.

As a specific example, consider the "coat-hook" problem: that of communicating a downward vertical point-load of magnitude F (the coat) to a vertical foundation arc (the wall), at distance d; see Figure (7.6). If the need to cope also with non-vertical loads is ignored, the problem is a two-dimensional one.

By Theorem (7.2), λ has value zero on the wall, and its downward component at the load point should be as large as possible, consistent with (7.32). The obvious invariance and scale features of the problem make it clear that λ has a constant direction, with magnitude depending only upon distance from the wall, and increasing linearly with this distance. That is, λ is of the form

$$\lambda(x) = x^1 v \qquad (7.54)$$

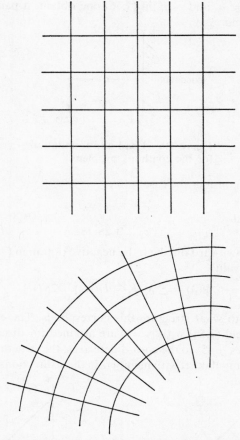

Figure 7.5 Examples of Hencky–Prandtl nets: rectangular nets and circular fans. See also the conjugate families of equiangular spirals in Figure (7.8)

for some fixed vector $v = (v_1, v_2)$, so that

$$E = \begin{pmatrix} v_1 & \tfrac{1}{2}v_2 \\ \tfrac{1}{2}v_2 & . \end{pmatrix}. \tag{7.55}$$

Since the structure will certainly contain both tension and compression members, both bounds will be attained in (7.32), and E will have α_T and $-\alpha_C$ as eigenvalues: i.e. these will be the two roots of the quadratic

$$\kappa(\kappa - v_1) = (v_2/2)^2. \tag{7.56}$$

Substituting $\kappa = \alpha_T$ and $-\alpha_C$ in (7.56) one obtains a pair of determining equations for v_1 and v_2.

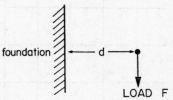

Figure 7.6 Load and foundation for the coathook problem

For simplicity, consider the case

$$\alpha_T = \alpha_C = \alpha,$$

so that $v_1 = 0$, and

$$v_2 = \pm 2\alpha. \tag{7.57}$$

Since λ is to be downward directed, the negative option in (7.57) is taken, and so the maximum value

$$\Phi(\lambda) = \sum \lambda_i \cdot F_i = Fd|v_2| = 2\alpha Fd \tag{7.58}$$

is achieved.

Consider now to what structure this corresponds. The extreme values of $D\lambda/Dx$ for the field (7.54) with $v_1 = 0$ are attained for directions inclined at the constant angles $\pm 45°$ to the co-ordinate axes, the structure must thus take the form of a rectangular net with members in these directions; see Figure (7.7).

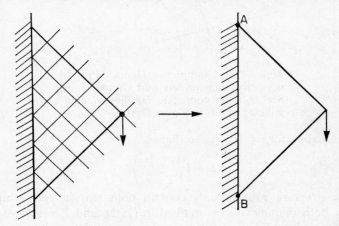

Figure 7.7 The Hencky–Prandtl net and its simplification for the coat-hook problem

One guesses that the portion of the net extending outside the two heavily drawn members will not be used: one sees then from statical considerations at the boundary that the members inside these two will carry no load, and so can also be deleted.

The presumption is, then, that the optimal framework will simplify to the two-member structure of Figure (7.7). This is indeed optimal. Resolution of forces at the load-point shows that the members will both be carrying longitudinal forces of $F/\sqrt{2}$ (the upper in tension, the lower in compression) and so will cost $\alpha F/\sqrt{2}$ per unit length. But the total length of member is $(2\sqrt{2})d$, so the total cost is $2\alpha Fd$. This upper bound for the minimal cost coincides with the lower bound (7.58), so the optimal structure has indeed been achieved.

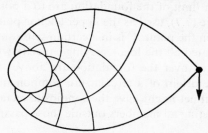

Figure 7.8 Communication of a point-load to a circular foundation arc; the design of a crank. The Hencky–Prandtl net consists of two families of equi-angular spirals

If the foundation arc is of circular rather than linear cross-section, then the optimal net is no longer rectangular, but made up of equiangular spirals (see Exercise (7.12)). This solves the problem of transferring a point-load to the circumference of a circle, i.e. of designing a crank.

The net extends to infinity, but in fact the only part needed for the optimal design is the part bounded by the two spirals passing through the load-point: heavily drawn in Figure (7.8). Since the boundaries of the net are now curved, we cannot dispense with the inner members; these will be required to equilibrate stresses normal to the boundary. However, they will carry relatively light loads, except at points where the curvature of the boundary is relatively large. The curvature will be greatest near the axle, and so the structure will be heavier and denser there. Indeed if the radius of the axle is small enough relative to the length of the crank-arm, then the bounding spirals will meet there and wrap around the axle to form a boss to the crank.

The continuous framework envisaged would have to be approximated, of course, by a discrete framework (with straight members) of small mesh.

Alternatively, if the crank were made in the solid, of a perfectly plastic material, the continuous framework theory is applicable. The solid takes the form of the framework (with appropriate α_T/α_C ratio) with density proportional to the density of matter in the optimal framework. For example, an ideal crank of a solid plastic material would have the form of Figure (7.8), the outer members, which carry the principal loads, being represented by ribs, and the inner members, which equilibrate stresses normal to the boundary, being represented by a web whose thickness increases towards the axle-end of the crank. If the axle is of small radius, the outer ribs wrap around it to form a boss.

Suppose one modifies the original coat-hook problem of Figure (7.6) by restricting the upper limit of the foundation arc to a point A' which is below the point A of Figure (7.7), used as the upper anchor-point of the unrestricted optimal design. Then the point A' is in a certain sense singular, because upon it must be concentrated all the load-carrying capacity which could otherwise have been distributed over the foundation-arc above A'. One guesses then that it might be the origin of a circular-fan component of the optimal net, and that the optimal net might have the composite rectangular/fan character of Figure (7.9). Deleting all members outside those passing through the load-

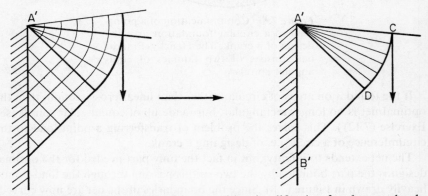

Figure 7.9 The Hencky–Prandtl net and its simplification for the coat-hook problem in the case where the foundation arc does not reach as high as the point A of Figure (7.7)

point and deleting all internal members which meet a straight boundary (and so carry no load), one derives the simplified framework on the right of Figure (7.9). This consists of three straight members, a circular-arc boundary member CD, and a continuum of radial members in tension between $A'C$ and $A'D$. This framework is statically determinate and optimal (see Exercise (7.4)).

A further restriction in the same direction would be to take the foundation arc as consisting just of a pair of points, not necessarily coinciding with the two foundation points actually used in either of our previous constructions. This is the design of a cantilever framework. A graphical method which gives an approximately Michell structure is the following, illustrated in Figure (7.10). Assume that the optimal framework will be locally fan-shaped in the

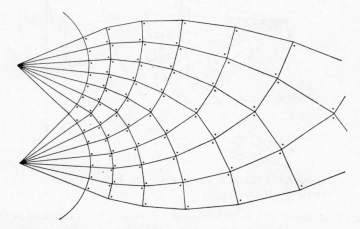

Figure 7.10   The coat-hook problem with two assigned foundation points. Graphical design of a cantilever structure. Right angles are marked with a dot

neighbourhood of the two foundation points: draw then, upon each as centre, a circular arc of such radius that the two arcs meet at right-angles. The net is then continued out from this compound arc by choosing a number of equally spaced points along each circular arc (the significant point is not the equality in *spacing*, but equality in increment of the *angle ϕ*) and from each continuing the radius vector. Where the radii of different systems meet, the net is continued by observing the right-angle requirements marked by a dot in Figure (7.10).

The net thus generated is a discrete approximation (with straight members) to a Hencky–Prandtl net. Complete orthogonality at all nodes is, of course, not possible, with straight members and a non-rectangular net. However, incorporation of the right-angles marked supplies an approximate orthogonality, and the fact, easily proved, that each quadrilateral mesh of the net has the same set of corner angles, supplies an approximate equivalent of the constant ϕ-increment property.

The net is then completed, and adapted to the case of a given load-point by drawing in the radial members from the two foundation points, approximating the elements of arc by chords, and deleting all members outside the generators passing through the load-point. The result is a statically determinate structure, pictured in Figure (7.11), in which the loads (and so cross-

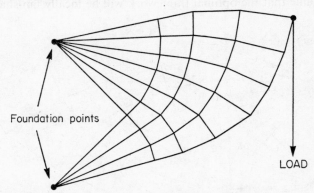

Figure 7.11 Simplification of Figure 7.10 to the actual cantilever structure

sections of members) can be determined recursively from the load-point. The constant geometry of the nodes eases this calculation greatly.

This construction has not been justified in detail, but the net is obviously approximately Hencky–Prandtl, and the choice of a local fan-structure at each foundation point seems reasonable. The net described is, in fact, approximately optimal if $\alpha_C = \alpha_T$, and the load is parallel to the line of the foundation points. The same intuitive/graphical considerations are just as helpful for more complicated structures.

Exercises

(7.11) Consider the coat-hook problem with an unlimited wall, but with $\alpha_T \neq \alpha_C$. Show that the optimal structure is still a rectangular net, reducing to two members, and determine it.

(7.12) Consider the crank problem of Figure (7.8), i.e. the case where a point-load at B perpendicular to AB is to be transferred to the circumference of a circle centred on A (see Figure (7.12)). We suppose that $\alpha_C = \alpha_T$. Consider a field $\lambda(x)$ whose direction is always perpendicular to the radius vector from A, and whose magnitude is a function $h(r)$ of the length r of this radius vector. Determine $h(r)$ so that both bounds are attained in (7.32). Show that the lines of the net corresponding to the λ-field thus determined are equi-angular spirals with centre A.

(7.13) Calculate the λ-field corresponding to a fan-net with radii in tension and arcs in compression, in the case $\alpha_C = \alpha_T$.

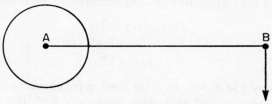

Figure 7.12

(7.14) Calculate the λ-field corresponding to the configuration of Figure (7.9), and hence, by comparing the corresponding values of the criterion function for primal and dual, demonstrate the optimality of this configuration.

7.4 THE MAXIMUM TRANSFORM

There is one very special case of a constrained maximization problem in which the Lagrangian results take a simple and striking form. Suppose the constraint $g(x) = b$ reduces simply to $x = b$. Then $U(b) = f(b)$, and the assertion that $U(b) = \hat{U}(b)$ under the conditions of Theorem (3.9) can be expressed in the following form:

Theorem (7.5)
Define the maximum transform of a function $f(x)$ on a set $\mathscr{X}$ by

$$F(y) = \max_{x \in \mathscr{X}} [f(x) - y^T x]. \tag{7.59}$$

This relation has the inversion

$$f(x) = \min_y [F(y) + y^T x] \tag{7.60}$$

at a point x if the following conditions are fulfilled:
  (i) *$\mathscr{X}$ is convex and f concave in $\mathscr{X}$,*
  (ii) *x is finite,*
  (iii) *either x is interior to $\mathscr{X}$ and f finite in a neighbourhood of x, or f has finite derivative at x in directions from x into $\mathscr{X}$.*

Thus, the nonlinear transform (7.59) has the simple and symmetric inversion (7.60), subject to the conditions stated. Relation (7.60) expresses f as the *minimum transform* of F. The pair of relations (7.59) and (7.60) are reminiscent of a Fourier transformation and its inverse, and can indeed be regarded as a limit case of such a relation-pair; see Section (8.6).

Although the inversion (7.60) follows as a special case of Lagrangian theory, the case is so special that we gain clarity by recapitulation and specialization of the argument. From the definition (7.59) it follows that

$$F(y) \geqslant f(x) - y^T x \qquad (7.61)$$

or

$$f(x) \leqslant F(y) + y^T x, \qquad (7.62)$$

for any y, and any x in $\mathscr{X}$. On the other hand, it is known from the concavity and regularity assumptions that there exists a y such that $f(\xi) - y^T \xi$ is maximal for ξ in $\mathscr{X}$ at $\xi = x$, so that

$$f(x) = \max [f(\xi) - y^T \xi] + y^T x$$
$$= F(y) + y^T x \qquad (7.63)$$

for this particular y value. Relations (7.62) and (7.63) imply (7.60).

The solutions of some of the earlier problems can be regarded as solutions in terms of maximum transforms. For example, for the additive allocation problem of (6.2)

$$U(b) = \min_y \left[\sum_1^n F_j(y) + yb \right], \qquad (7.64)$$

where y is a scalar,

$$F_j(y) = \max_{x \geqslant 0} [f_j(x) - yx], \qquad (7.65)$$

and the minimization in (7.64) is over all y, or only over positive y, according as the constraint is $\sum x_k = b$, or $\sum x_k \leqslant b$.

Note also the occurrence of $\max_u [h(u) - yu]$ in the dam problem of Section (6.3); effectively, the occurrence of a maximum transform. This type of application, in control theory, is discussed further in the next section.

A geometric interpretation of the maximum transform and its inversion has virtually been given in Section (3.1); also an interpretation of what happens in the case when $f(x)$ is non-concave, and one may have

$$f(x) < \min_y [F(y) + y^T x].$$

An economic interpretation of the inversion can also be given, somewhat on the lines of the interpretation of the dual of an allocation problem. Indeed, one views the problem as a degenerate allocation problem. Suppose that a supplier can purchase goods at a fixed (vector) cost price y, and that by selling a (vector) amount of goods x he can obtain a gross return $f(x)$.

His net profit is then $f(x) - y^T x$, and he will choose x, the amount handled, to maximize this. His profit, as a function of cost price, will then be

$$F(y) = \max_{x \geq 0} [f(x) - y^T x]. \tag{7.66}$$

$F(y)$ can be regarded as the value of the supplier's services. Inequality (7.62) states that the value to the public of goods x is not greater than the sum of values of goods and services, for any cost price y. Inversion (7.60) states that, granted the convexity assumptions etc., there is a price y for which equality holds in (7.61), this price necessarily minimizing the combined value of goods and services, when an amount x of goods is being handled.

It is known from the arguments of Sections (3.1) and (3.8) that, if $\mathscr{X}, f$ are such that there is no value of y yielding equality in (7.62), then

$$\hat{f}(x) = \min_{y} [F(y) + y^T x] \tag{7.67}$$

still represents the average return that can be achieved (using the language of the last example) with a randomized solution. Specifically, a distribution $\{p_j\}$ can be found at most $n+1$ points x^j (if x is an n-vector) such that

$$\sum p_j x^j = x \tag{7.68}$$

and

$$\sum p_j f(x^j) = \hat{f}(x). \tag{7.69}$$

That is, the supplier can handle an amount x^j a fraction p_j of the time, so handling an average amount x, and achieve an average gross return of $\hat{f}(x)$.

Exercises

(7.15) Calculate the maximum transform and verify the inversion (7.60) for the following cases, x being scalar unless otherwise indicated.

(i) $f(x) = \begin{cases} \log(x-a) & (x > a), \\ -\infty & (x \leq a). \end{cases}$

(ii) $f(x) = -\frac{1}{2}(x-a)^2$, with x unrestricted, and for the vector generalization of this case.

(iii) As in (ii), but with x restricted to an interval of the real axis.

(iv) $f(x) = -\frac{1}{2}(a + |x|)^2$.

(v) $f(x) = -|x|^\nu$, where $\nu \geq 1$.

(vi) $f(x)$ concave and piecewise linear.

(vii) $f(x) = c^T x$, with $\mathscr{X}$ the hypercube:

$$0 \leq x_k \leq M \quad (k = 1, 2, \ldots, n)$$

(7.16) Suppose that $f(x) = c^{\mathrm{T}} x$ and $\mathscr{X} = R_+{}^n$. Show that

$$F(y) = \begin{cases} 0 & \text{if } y \geqslant c, \\ +\infty & \text{otherwise,} \end{cases}$$

and verify the inversion.

(7.17) Suppose that $f(x) = \min_{1 \leqslant k \leqslant n} (c_k x_k)$ where the c_k are all positive, and that $\mathscr{X} = R_+{}^n$. Show that

$$F(y) = \begin{cases} 0 & \text{if } y \geqslant 0 \text{ and } \sum y_k / c_k \geqslant 1, \\ +\infty & \text{otherwise,} \end{cases}$$

and verify the inversion.

(7.18) Show, by appeal to Exercise (7.17) if desired, that $f(x) = \min_j f_j(x)$ does not necessarily imply $F(y) = \min_j F_j(y)$.

(7.19) Show that $F(y)$ is convex.

(7.20) Show that $\hat{f}(x)$, defined in (7.67), is concave. Show that there is no concave function $\phi(x)$ such that $\phi(x) \geqslant f(x)$ in $\mathscr{X}$ and that $\phi(x) < \hat{f}(x)$ for any x in the convex hull of $\mathscr{X}$. That is, $\hat{f}(x)$ is the least concave function majorizing f in $\mathscr{X}$.

(7.21) Suppose $\mathscr{X}$ is a discrete point set. Show that $\hat{f}$ is piecewise linear. So, $\hat{f}$ can be regarded as providing a linear interpolation of f between its values in $\mathscr{X}$.

(7.22) Show from the definition (7.59) that, provided all derivatives exist and x is an interior point of $\mathscr{X}$, then $F_y = -x^{\mathrm{T}}$ and $F_{yy} = -f_{xx}{}^{-1}$. Here the values of y and x are understood to be corresponding values; i.e. x is the value for which the maximum is achieved in (7.59) at the y-value considered.

(7.23) Relate the maximum transform to the classic *Legendre trans-formation*:

$$\mathscr{L}(p) = f(x) - p^{\mathrm{T}} x$$

where

$$p_k = \frac{\partial f}{\partial x_k}.$$

(7.24) Consider the relationship

$$\phi(x) = \max_u [h(u) + \psi(Ax + Bu)].$$

Show that this becomes, in terms of maximum transforms,

$$\Phi(A^{\mathrm{T}} y) = H(-B^{\mathrm{T}} y) + \Psi(y).$$

Hence consider the maximization of $\sum_0^{N-1} h(u_t) + k(x_N)$ subject to prescribed x_0 and to

$$x_{t+1} = Ax_t + Bu_t \quad (t \geqslant 0).$$

(See Section (7.5).)

(7.25) (Bellman). Consider the functional equation for $f(x)$:

$$f(x) = a(x) + \max_\xi [f(\xi) b(x - \xi)].$$

Show that this has the formal solution

$$f(x) = \min_y \left[\frac{e^{xy} \tilde{a}(y)}{1 - \tilde{b}(y)} \right],$$

where

$$\tilde{a}(y) = \max_x [a(x) e^{-xy}], \quad \text{etc.}$$

and determine conditions for its validity.

7.5 APPLICATION OF THE MAXIMUM TRANSFORM TO LINEAR CONTROL

Some special cases of the control problem of Section (7.1) are amenable to solution by maximum transforms. This approach, when it works, is completely equivalent to use of the maximum principle (i.e. of strong Lagrangian methods). However, explicit use of the transform is interesting and gives a neat treatment.

It will be supposed that the dynamic relation has the linear form (7.8), and that the utility function to be maximized is of the form

$$f(x, u) = \sum_{t=0}^{N-1} h_t(u_t) + k(x_N). \tag{7.70}$$

That is, it has the additive form (7.4), with the additional specialization that the value of the state variable affects utility only at termination: costs and utility incurred during the course of the process depend only on the control variable.

In order that f may be concave, k and the h_t are assumed concave. It is further assumed that any restrictions on the value of (x, u) can be built into f (so that this tends to $-\infty$ as (x, u) tends towards a forbidden value).

It is most convenient to put the problem into a recursive form. Let

$$V_t(x_t) = \max_{u_t \ldots u_{N-1}} \left[\sum_{s=t}^{N-1} h_s(u_s) + k(x_N) \right], \tag{7.71}$$

the maximization being subject to (7.8). Then, plainly,

$$V_N(x) = k(x),$$

and

$$V_t(x) = \max_u [h_t(u) + V_{t+1}(Ax + Bu)] \quad (t < N). \tag{7.72}$$

Recursion (7.72) is the *dynamic programming equation*: if x is given the value x_t then the maximizing u in (7.72) is just the optimal value of u_t.

The functions $V_N = k$ and h_t (all t) are concave. Then a recursive argument, using (7.72), establishes the fact that the functions $V_t(x)$ are also concave, in x. The maximum transform is now defined

$$\tilde{V}_t(y) = \max_x [V_t(x) - y^T x] \tag{7.73}$$

and correspondingly for $\tilde{k}, \tilde{h}_t$. Relation (7.72) can then be rewritten

$$\tilde{V}_t(A^T y) = \tilde{h}_t(-B^T y) + \tilde{V}_{t+1}(y), \tag{7.74}$$

i.e. as a recursion which is free of the nonlinear operator, max. If A is non-singular (see Exercise (7.31) for consideration of the singular case), then (7.74) has the immediate solution

$$\tilde{V}_t(y) = \sum_{s=t}^{N-1} \tilde{h}_s(-B^T(A^T)^{t-s} y) + \tilde{k}((A^T)^{t-N} y). \tag{7.75}$$

$V_t(x)$ is then determined from the inversion

$$V_t(x) = \min_y [\tilde{V}_t(y) + y^T x], \tag{7.76}$$

and the optimal u_t by the maximization of $h_t(u) + V_{t+1}(Ax_t + Bu)$.

As an example, consider a very simple economic problem, concerning the allocation of resources between consumption and investment. Let all variables be scalar; let x_t denote the amount of resources available at the beginning of the tth stage, and let u_t denote the amount consumed, giving a utility $h_t(u_t)$. Suppose that

$$x_{t+1} = \alpha(x_t - u_t), \tag{7.77}$$

so that the remaining resources $x_t - u_t$, when reinvested, produce an initial capital $\alpha(x_t - u_t)$ for the beginning of the next stage. The multiplication factor α will normally be greater than unity, although this is not necessary.

Assume also that the "economic plan" terminates at time N, and the residual resources x_N then have a terminal utility $k(x_N)$.

Solution (7.75) becomes, in this simple case,

$$\tilde{V}_t(y) = \sum_{s=t}^{N-1} \tilde{h}_t(\alpha^{t-s} y) + \tilde{k}(\alpha^{t-N} y). \tag{7.78}$$

For example, suppose that

$$h_t(u) = a \log(u - u^*),$$
$$k(x) = b \log(x - x_N^*), \tag{7.79}$$

where the logarithms are assigned value $-\infty$ if their argument is negative. Thus, u^* is the subsistence level of consumption, below which consumption must not fall, and x_N^* is the minimal acceptable level of terminal resources.

We have then

$$\tilde{h}_t(y) = \begin{cases} +\infty & (y \leqslant 0), \\ a \log(a/y) - a - u^* y & (y > 0). \end{cases} \tag{7.80}$$

Evaluation of (7.78) and its inversion by (7.76) then yield

$$V_t(x) = a\tau \log a + b \log b + [\tfrac{1}{2} a\tau(\tau - 1) + b\tau] \log \alpha$$
$$+ (a\tau + b) \log \left(\frac{x - x_t^*}{a\tau + b} \right). \tag{7.81}$$

Here

$$\tau = N - t, \tag{7.82}$$

the "time-to-go" before the end of the plan; the logarithm is again assigned value $-\infty$ for negative argument, and

$$x_t^* = \alpha^{t-N} x_N^* + \frac{1 - \alpha^{t-N}}{1 - \alpha^{-1}} u^*. \tag{7.83}$$

The quantity x_t^* plainly has the interpretation: the minimal amount of resources at time t which will enable one to keep future consumption above subsistence level and final resources above the target amount.

We find from (7.72), (7.81) that the optimal consumption at time t is

$$u_t = \frac{(a\tau + b) u^* + a(x_t - x_t^*)}{a(\tau + 1) + b}; \tag{7.84}$$

provided $x_t \geqslant x_t^*$. That is, u_t is an averaged value of the subsistence level u^* and the "surplus" $x_t - x_t^*$, but approaches u^* as τ increases. That is, even with a surplus, one tends to live at the subsistence level until near the end of the plan. This rather unrealistic feature can be removed if an incentive to short-term consumption is built into the model (see Exercise (7.32)).

The restrictions $x_t \geqslant 0$ (or, equivalently, $u_t \leqslant x_t$) have not been imposed. To impose them is unnecessary, because the stronger condition $x_t \geqslant x_t^*$ is already implicit in the problem.

Exercises

(7.26) Work through the reinvestment example just considered by the Pontryagin method of Section (7.1).

(7.27) Consider the reinvestment problem with a utility function described by

$$h_t(u) = \begin{cases} a(u - u^*) & (u \geqslant u^*), \\ -\infty & (u < u^*), \end{cases}$$

and correspondingly for k.

(7.28) Consider the reinvestment problem for the cases of time-dependent h, and vector x, u.

(7.29) Prove concavity of the $V_t(x)$, given concavity of k and the h_t.

(7.30) Verify the solution (7.81) directly from (7.72).

(7.31) $V_t(x)$ is a function only of the linearly independent components of $A^{N-t}x$, and so is degenerate if A is singular. For example, if $A = 0$ then $V_t(x)$ is independent of x for $t < N$, and its maximum transform is

$$\tilde{V}_t(y) = \begin{cases} V_t(x) & (y = 0), \\ +\infty & (y \neq 0), \end{cases}$$

which explains why the recurrence (7.74) does not then determine V_t for all y.

(7.32) Suppose that one *discounts the future*, by taking

$$\sum_0^{N-1} \beta^t h_t(u_t) + \beta^N k(x_N)$$

as the quantity to be maximized, instead of (7.70), and defining the maximal future utility by

$$V_t(x_t) = \max_{u_t \ldots u_{N-1}} \left[\sum_{s=t}^{N-1} \beta^{s-t} h_s(u_s) + \beta^{N-t} k(x_N) \right]$$

instead of by (7.71). Here β is a discount factor, lying between 0 and 1. Calculate the solution for $V_t(x)$ in the case (7.79). Show that if $\beta < 1$ and $\alpha > 1$ then $V_t(x)$ has a limit as $N \to \infty$, and that the rule relating u_t and x_t in these circumstances is

$$u = u^* + (1 - \beta)(x - x^*),$$

where

$$x^* = \frac{u^*}{1 - \alpha^{-1}}.$$

Interpret x^*. (The legitimacy of discounting, in some objective sense, is still a matter of discussion. However, at the level of the individual one can see a direct justification for it. Suppose a person in life has a probability β of surviving the next stage, independent of previous events. Then the probability that he will still be in life ν stages hences is proportional to β^ν, and such a weighting term must be introduced in the expression above for the *expected* utility. As is seen from the solution of this exercise, propensity to consume increases with the probability, $1 - \beta$, of death.)

CHAPTER 8

Nonlinear Constraints, and Stochastic Effects

This chapter collects a number of miscellaneous but important topics. The last four chapters have all been concerned with the case of linear equality constraints: Section (8.1) touches on other types of constraint amenable to strong Lagrangian techniques. In some sense the obvious nonlinear extensions are just those which are formally reducible to the linear case, as is noted in Section (8.3).

Sections (8.4) and (8.5) deal rather passingly with what is known as stochastic linear programming: a constrained maximization problem in which there are statistical uncertainties either in the return (criterion) function or in the constraints.

However, Sections (8.6) and (8.7) are concerned with statistical effects of a deeper character. Suppose that x is a random variable, taking values in $\mathcal{X}$, and that, instead of considering the maximum of $f(x)$ under the constraint $g(x) = b$ (say), we consider the *probability distribution* of $f(x)$ conditional on the event $g(x) = b$. This generalized problem exhibits Lagrangian techniques in a much more general light, and many points acquire an added significance. For example, primal and dual formulations generalize in a remarkable manner, and the special character of linear constraints emerges clearly. There are several practical situations in which this distributional generalization of the constrained problem is of interest: some equilibrium problems of statistical mechanics are considered specifically.

The chapter falls into four virtually independent groups of sections: (8.1)–(8.2), (8.3), (8.4)–(8.5) and (8.6)–(8.7).

8.1 NONLINEAR CONSTRAINTS

If the strong Lagrangian principle is to be valid for a family of problems $P(b)$, the essential requirements are that $\mathcal{B}$ should be convex, and $U(b)$ concave on $\mathcal{B}$. It is then impossible to improve a solution by mixing: to increase the average value of $f(x)$ by taking a distribution of x-values, consistent with fulfilment of the constraint on average.

Theorem (3.10) represents the most general set of conditions derived so far which imply these convexity properties. The conditions of Theorem (3.10) are by no means necessary, as will be seen below. On the other hand, it does not seem possible to weaken them in a radical fashion, and one is forced to the conclusion that there are indeed many cases of practical interest where the Lagrangian principle is simply not valid in the strong form, although it may be in the weak.

Consider first the case of equality constraints: $g(x) = b$. Theorem (3.10) then requires linearity of $g(x)$, in addition to concavity of $f(x)$ in $\mathcal{X}$. These conditions are severe, and obviously not necessary. For example, it will be sufficient if there is a one-to-one transformation of x for which the transformed $f, g, \mathcal{X}$ have these properties. Again, if it is required only that the single problem $P(b)$ be soluble by Lagrangian methods, then a supporting hyperplane to the graph of U need exist only at the single point b, and this may exist without U's being concave.

It is particularly in the control examples that one would wish to be able to work with nonlinear equality constraints: one would like to allow the function $r_t(x_t, u_t)$ of (7.1) to be nonlinear in x_t and u_t. Various mild extensions have been made in this direction (see Exercises (8.3)–(8.5)). An interesting point that emerges is that it is more natural to consider the relation of x_{t+1} to x_t, u_t as a set-to-set mapping, rather than as a point-to-point mapping given by a functional relation (7.1). This is an observation that is followed up to some extent in the example of the next section.

In general, however, effort in this direction does not seem particularly well spent, for two reasons. One is that the strong Lagrangian principle is definitely invalid for certain cases of practical interest (cf. Exercise (8.6)). The other is that already mentioned: that validity of the Pontryagin maximum principle in the more natural case of continuous time does not basically turn on convexity conditions at all.

Theorem (3.10) certainly allows a measure of nonlinearity to be introduced into $g(x)$ if the constraints are not equalities. For example, consider again the nonlinear allocation problem introduced in Section (2.1), of maximizing $f(x)$ in $x \geqslant 0$, subject to

$$g(x) \leqslant b. \tag{8.1}$$

The cone $\mathscr{C}$ of Theorem (3.10) is thus the positive orthant, $R_+{}^m$, and $\mathscr{C}$-convexity corresponds to ordinary convexity of the elements of $g(x)$.

Note, in passing, that to take account of the fuel limitation condition (7.10) of the "rocket" problem by a Lagrangian multiplier was indeed justified. The function $\sum |u_t|$ is convex in (x, u), as Theorem (3.10) will require.

Returning to the allocation problem, one sees that Theorem (3.10) will justify the conclusion that there exists a set of effective prices, y, such that the solution $\bar{x}$ of the problem maximizes $f(x) - y^T g(x)$ freely in $x \geqslant 0$, provided $f(x)$ is assumed concave and $g(x)$ convex (with supplementary regularity conditions if y is to be finite). The assumption of concave $f(x)$ is reasonable; the familiar concept of decreasing rate of return with increase in activity. The assumed convexity of $g(x)$ might conceivably be justified on the same grounds: that production output x should be concave in production input $g(x)$. It is implied that the convex mapping $x \rightarrow g(x)$ can be assigned some kind of inverse, which is then necessarily concave: the point needs discussion, but is left intuitive.) However, it would seem more realistic to postulate the opposite conclusion: that economies of scale would lead to more efficient use of resources as production intensities x are increased, so that $g(x)$ should be *concave*.

The case of concave $g(x)$, when economies of scale constitute the predominant effect, seems the more realistic one economically, even though it is not covered by the Lagrangian theorems hitherto developed. Certainly the character of the optimal point can be very different if $g(x)$ is concave, cf. Exercise (2.12), and there are cases for which the optimal point definitely cannot be located by Lagrangian methods at all; cf. Exercises (8.1), (8.2). One imagines that Lagrangian methods would still be valid under the reasonable assumption that $f(x)$ is "more concave" than $g(x)$ in some appropriate sense, but this is a point which has not been cleared up.

Exercises

(8.1) Consider the maximization of $f(x) = x_1 + x_2$ in $x \geqslant 0$, subject to the concave constraint

$$\sqrt{x_1} + \sqrt{x_2} \leqslant b.$$

Show that, by searching for stationary points of the Lagrangian form, one does not locate the maximum (achieved at $x = (b^2, 0)$ or $(0, b^2)$ but rather the *minimum* of f (achieved at $x = 0$, or at $(b^2/4, b^2/4)$, if equality is demanded in the constraint).

(8.2) Consider the maximization of $f(x) = c^T x$ in $x \geqslant 0$ subject to the single constraint $\sum h_k(x_k) \leqslant b$, where

$$h_k(x_k) = \begin{cases} 0 & (x_k = 0), \\ \alpha_k + \beta_k x_k & (x_k > 0) \end{cases}$$

and the α_k and β_k are positive constants. This can be regarded as an allocation problem with a linear return function, a single critical resource (say, capital) and production costs which are linear but for a setting-up cost, α_k. The setting-up component makes the production costs concave functions of intensity.

Show that the solution to this problem does not satisfy a strong Lagrangian principle.

(8.3) The control problem of Section (7.1) in the case of a time-additive utility (7.4) can always be transformed so that the components $h_t(x_t, u_t)$ in the criterion function are zero, i.e. so that the problem is a *terminal problem*. To do this, introduce a supplementary component x_t^0 in the state variable, satisfying the recursion

$$x_{t+1}^0 = x_t^0 + h_t(x_t, u_t).$$

This relation supplements (7.1). The total utility (7.4) can then be written $x_N^0 + k(x_N)$: a function purely of the terminal value of the enlarged state variable.

Suppose this normalization already performed, so that the recursion is just (7.1), and the criterion function is $k(x_N)$. Denote the expression $x_{t+1} - r_t(x_t, u_t)$ by b_t ($t = 0, 1, ..., N-1$). Show that $U(b)$ has the required concavity properties, if, given any two pairs of sequences (x', u') and (x'', u'') in the permitted set $\mathscr{W}$, with associated b vectors b', b'' and any p, q ($p, q \geqslant 0, p + q = 1$), one can find an (x, u) sequence in $\mathscr{W}$ such that $b = pb' + qb''$ and $k(x_N) \geqslant pk(x_N') + qk(x_N'')$.

(8.4) Consider the discrete control problem in the normalized form of Exercise (8.3). Suppose that $\mathscr{W}$ has the specification: x_t must lie in a convex set $\mathscr{X}_t$, and u_t in a set $\mathscr{U}_t$ ($t = 0, 1, 2, ...$). Suppose $k(x_N)$ concave, and that for any p, q (usual conventions) and for any x_t', x_t'' in $\mathscr{X}_t$ and u_t', u_t'' in $\mathscr{U}_t$ there exists a u_t in $\mathscr{U}_t$ such that

$$r_t(px_t' + qx_t'', u_t) = pr_t(x_t', u_t') + qr_t(x_t'', u_t'') \quad (t = 0, 1, ..., N-1). \quad (8.2)$$

Show that $U(b)$ has then the required concavity properties.

(8.5) The convexity-type assumption (8.2) can be relaxed, at the expense of stronger conditions in other directions. Consider the following modification of the conditions of the second last sentence of Exercise (8.4). Suppose there exists a cone $\mathscr{C}$ such that, if x_t belongs to $\mathscr{X}_t$ and z to $\mathscr{C}$ then $x_t + z$ also belongs to $\mathscr{X}_t$. Suppose that r_t is $\mathscr{C}$-increasing in the sense that

$$r_t(x_t + z, u_t) - r_t(x_t, u_t)$$

belongs to $\mathscr{C}$ for z in $\mathscr{C}$, and that k is concave, and also $\mathscr{C}$-increasing in the sense that $k(x_N + z) \geqslant k(x_N)$. Suppose finally that (8.2) is replaced by: there exists a u_t in $\mathscr{U}_t$ such that the left-hand member of (8.2) exceeds the right-hand member by an element of $\mathscr{C}$. Show that $U(b)$ then has the required concavity properties.

(8.6) Consider the cross-current extraction problem of Section (2.5), and suppose the function $h(x)$ of (2.29) linear and increasing. Suppose that wash-water has a unit cost of λ, so that, if x_N is the concentration of solute in the stream entering the last extractor, u_N will be chosen so as to minimize $x_{N+1} + \lambda u_N$. Consider the minimized value of this quantity as a function of x_N; $U(x_N)$. Show that U is concave, although this is a minimization problem, for which one would hope for convex cost functions. This concavity shows that one could not use a strong Lagrangian principle in optimizing the final stage of extraction (still less in optimizing all stages) although the treatment of Section (2.5) shows how useful the Lagrangian principle still is, in its weak form.

8.2 ECONOMIC DEVELOPMENT PROGRAMMES WITH NONLINEAR RECURSIONS

Consider now yet another version of the problem considered in Section (5.1): the optimization of a dynamic economy. The treatment (due to Gale, 1968) is interesting, in that it successfully applies Lagrangian methods, while allowing nonlinearity in the relation between the variables of consecutive stages. The device employed may seem to side-step the issue of nonlinearity, but it nevertheless leads to fairly specific conclusions.

It will be assumed that $x(t)$ represents the vector of amounts of resources available at time t (not, as in Section (5.1), the intensities of activities at that time). The variable $u(t)$, which represents the element of choice available in the passage from $x(t)$ to $x(t+1)$, then enters in two ways. Firstly, part of $x(t)$ can be diverted for immediate consumption, and then, the remaining part can be allocated over the processes of production in various ways. It is desired to optimize both of these compound decisions.

Let the vector amounts consumed and reinvested at time t be denoted $v(t)$ and $w(t)$, so that

$$v(t) + w(t) \leqslant x(t). \tag{8.3}$$

The relationship between $w(t)$ and $x(t+1)$, the amounts reinvested and then produced, can take various forms according as the allocation of reinvestment is varied. However, it is not necessary to assume a linear relationship, nor, indeed, to put the process in the form of a functional relationship, such as (7.1), at all. Rather, it is better to assume that $(w(t), x(t+1))$ can jointly take any value-pair in a convex set $\mathcal{T}(t)$, known as the *technology* at time t. The term "technology" is appropriate, in that, by specifying $\mathcal{T}(t)$, we specify what range of $x(t+1)$ could actually be produced from a given $w(t)$ at time t.

The approach is now to regard $\mathcal{T}(0) \times \mathcal{T}(1) \times \ldots \times \mathcal{T}(N-1) = \pi$ (say) as the specification of the basic region of variation (corresponding to $\mathcal{X}$ in the general formulation) and to take account of restrictions (8.3) by Lagrangian multipliers. That is, the nonlinear part of the $x(t) \to x(t+1)$ relationship has been sunk back into the specification of $\mathscr{X}$, leaving only the relatively trivial linear constraints (8.3) to be accounted for by Lagrangian methods.

This seems like an evasion of the problem; to some extent it is. However, even for this weakened version of the problem, Lagrangian methods furnish strong conclusions, and the notion of representing the recursion by a set rather than by a family of functional relationships seems a valuable one.

Suppose then that we wish to maximize the utility function

$$f(x, v, w) = \sum_{0}^{N-1} h_t(v(t)) + k(x(N)) \qquad (8.4)$$

within the set π ($x(0)$ being specified), subject to constraints (8.3). Here $h_t(v(t))$ represents the utility derived from consumption $v(t)$ at time t, and $k(x(N))$ the utility represented by an accumulation of resources $x(N)$ at the end of the programme. If these functions are supposed concave, then f is concave, the set π is convex, and the constraints (8.3) linear, so that one can appeal to the theorems of Section (3.6), and assert the existence of non-negative vectors $y(0), y(1), ..., y(N-1)$ such that (x, v, w) maximizes

$$L(x, v, w; y) = \sum_{t=0}^{N-1} [h_t(v(t)) + y(t)^{\mathrm{T}}(x(t) - v(t) - w(t))] + k(x(N)) \qquad (8.5)$$

in π. A number of conclusions then follow immediately.

Theorem (8.1)

Under the given assumptions on π, h_t and k, there exist non-negative vectors $y(0), y(1), ..., y(N-1)$ such that:

(i) *$(x(t+1), w(t))$ maximizes $y(t+1)^{\mathrm{T}} x(t+1) - y(t)^{\mathrm{T}} w(t)$ in $\mathscr{T}(t)$ $(t = 0, 1, ..., N-2)$ and $(x(N), w(N-1))$ maximizes $k(x(N)) - y(N-1)^{\mathrm{T}} w(N-1)$ in $\mathscr{T}(N-1)$;*

(ii) *$v(t)$ maximizes $h_t(v(t)) - y(t)^{\mathrm{T}} v(t)$;*

(iii) *$y(t)$ is interpretable as an effective marginal price for resources at stage t, in that, if (8.3) is modified to $v(t) + w(t) \leqslant x(t) + b(t)$ then*

$$y(t)^{\mathrm{T}} = \left[\frac{\partial U(b)}{\partial b(t)} \right]_{b(t) = 0} \qquad (8.6)$$

if this derivative exists, and otherwise obeys the inequalities analogous to (3.51).

All these assertions are formally immediate; only intepretation is necessary. Assertion (iii) states the now familiar price interpretation: total utility can be increased by an amount $y(t)^T b(t) + o(b(t))$ if $x(t)$ is increased to $x(t) + b(t)$ (with appropriate reservations if the derivative (8.6) does not exist). Assertion (i) then states that production is to be operated at the tth stage so as to maximize immediate profit, in terms of these prices. Assertion (ii) states that consumption at stage t is to be chosen at a level which maximizes net immediate utility, if consumption must be paid for at a price (utility rate) of $y(t)$.

Some of the consequences of the theorem are given in the exercises. They reflect the general tendency, noted already in Section (5.1) and at the end of Section (7.5), to hold consumption down until the end of the planning period approaches.

Exercises

(8.7) (Gale). Consider the case of a single good, and suppose that the technology takes the constant form of the set of points (x, w) such that $w \geqslant 0, x \leqslant G(w)$ where G is concave. Suppose further that the model is *productive*, in the sense that

$$G(w + \delta) - G(w) > \delta$$

for all $w \geqslant 0, \delta > 0$. Suppose also that $h_t(v)$ is independent of t, and both it and $k(x)$ are increasing. Show then that $v(t)$ is a non-decreasing function of t, in the optimal scheme.

(8.8) (Gale) Show, under the conditions of Exercise (8.7), and for the optimal scheme, that if $w(t)$ once begins to decrease then it continues to do so. Thus, inputs increase up to some point in time, and decrease thereafter.

(8.9) Relate the approach of this section to the formulation of Exercises (8.4) and (8.5).

8.3 THE REDUCTION OF PROGRAMMING PROBLEMS TO LINEAR FORM

Suppose that $f(x)$ is a function concave in $\mathscr{X}$, with finite directional derivatives. The set $(\mathscr{X}, f(\mathscr{X}) - R_+^1)$ in R^{n+1} will then certainly possess supporting hyperplanes at $(x', f(x'))$ for all x' in $\mathscr{X}$: suppose that these are

$$t = a_\nu^T x + b_\nu, \tag{8.7}$$

where ν traverses a set $\mathscr{N}$. That is

$$f(x) \leqslant a_\nu^T x + b_\nu \quad (x \in \mathscr{X}, \nu \in \mathscr{N}), \tag{8.8}$$

and for any x (in $\mathscr{X}$) there is a ν (in $\mathscr{N}$) for which equality holds in (8.8).

This fact implies the representation

$$f(x) = \min_\nu (a_\nu^T x + b_\nu) \tag{8.9}$$

of the concave function $f(x)$ in $\mathscr{X}$, sometimes termed a *quasilinear* representation of f. It is this representation which implies a linear formulation of many programming problems.

For example, consider the problem of maximizing $f(x)$ in $\mathscr{X}$, with no other constraint. If $\check{f}$ is the maximum attained then it follows from (8.8) that

$$\check{f} \leqslant a_\nu^{\mathrm{T}} x + b_\nu \quad (\nu \in \mathscr{N}) \tag{8.10}$$

for x equal to the maximizing value, and the problem can be rephrased as one of finding an x in $\mathscr{X}$ and an $\check{f}$ for which $\check{f}$ is maximal, consistent with (8.10). This is a constrained maximization problem in $(x, \check{f})$ involving only linear functions of these variables: see Exercise (8.10).

Suppose that the maximization is subject to the constraint

$$g(x) \leqslant 0. \tag{8.11}$$

The conditions of Theorem (3.10) will require that the elements $g_j(x)$ of $g(x)$ be convex, and so have a representation.

$$g_j(x) = \max_\nu (c_{j\nu}^{\mathrm{T}} x + d_{j\nu}), \tag{8.12}$$

for ν in some set $\mathscr{N}_j$, analogously to (8.9). The constraints (8.11) can thus be written in the form

$$c_{j\nu}^{\mathrm{T}} x + d_{j\nu} \leqslant 0 \quad (\nu \in \mathscr{N}_j; j = 1, 2, \ldots, m), \tag{8.13}$$

and the solution of the constrained problem will be the x in $\mathscr{X}$ for which $\check{f}$ is maximal, subject to the linear constraints (8.10) and (8.13) on x and $\check{f}$.

This "reduction" will seldom be at all useful, since the number of constraints in the linear version may be large, or infinite. However, it does demonstrate the basically linear character of strong Lagrangian theory. It also points the way to other "Lagrangian" theories. If a "convex" function is defined as one having the representation

$$f(x) = \max_\nu [\phi_\nu(x)] \tag{8.14}$$

for some family of functions ϕ_ν, then definition (8.14) implies a "supporting ϕ-surface" theorem, for the set $t \geqslant f(x)$, and so gives the beginning of a generalized Lagrangian theory.

However, the conventional concept of convexity (corresponding to linear ϕ_ν) seems to have a special status, in view of its relation with the concepts of closure under averaging, and of randomized solutions.

Exercises

(8.10) Consider the problem of maximizing a concave $f(x)$ unconstrainedly in $\mathscr{X}$. It has been shown that the maximum $\check{f}$ must satisfy (8.10), but not that the maximal $\bar{f}$ satisfying (8.10) is indeed the value of $f(x)$ for some x in $\mathscr{X}$. Demonstrate this point.

(8.11) Do the same for the problem with constraint (8.11).

(8.12) Construct a "linear reduction" of the general problem $P_{\mathscr{C}}(b)$, under appropriate convexity conditions.

8.4   STOCHASTIC LINEAR PROGRAMMING

There is considerable interest in programming situations which are affected by chance variation. One such case would be the linear allocation problem of (4.1)–(4.3) if A, b and c were to be regarded as random variables. The case when c is random (the case of uncertain returns) has already been considered in Section (6.2), but there are situations where chance enters the constraints as well. For example, the availabilities b of the various resources may be uncertain. In the diet problem of Section (4.4a) it may be that the stock feeds available are of variable composition, and that the concentrations of nutrients in them (specified by A) must be regarded as random variables.

The stochastic variants of a deterministic problem can take many forms, depending upon what random variables can be observed, and at what stage in the process they are observed. The introduction of a stochastic element can even convert a static process into a full-blown dynamic one, whose operation requires a whole sequence of decisions instead of a single one. For example, consider the flow problem of Section (5.3), for which optimization consisted in the determination of a single maximal-flow pattern, which is then held constantly. Suppose, however, that "traffic" travels at a random rate along the arcs, or enters the network at a random rate (the natural variation that will be inevitable, especially if "traffic" is made up of discrete units, such as consignments, or trains). There will then be accumulations, serious even if transitory, at points where local traffic momentarily exceeds capacity, and the optimal flow pattern will need constant revision in the light of the current state of traffic in the network.

Such problems are extremely difficult, and will certainly not be considered here.

Attention will be restricted to the case of a single allocation decision, when the intensity vector x must be chosen before the actual values of A, b and c are known. It must then be chosen solely on the knowledge of the distribution of these quantities, so that no guarantee can be given that the constraints

$$Ax \leqslant b \qquad\qquad (8.15)$$

will be fulfilled. One has then to accept the possibility that they may not be fulfilled, and to assign a cost or penalty in such cases. That is, rather than maximizing $c^T x$ subject to (8.15) and $x \geqslant 0$, one maximizes

$$E[H(c^T x) + K(Ax - b)] \qquad (8.16)$$

freely in $x \geqslant 0$. Here H is a utility function and K the negative of a loss function, having perhaps the general character of the functions graphed in Figure (8.1)

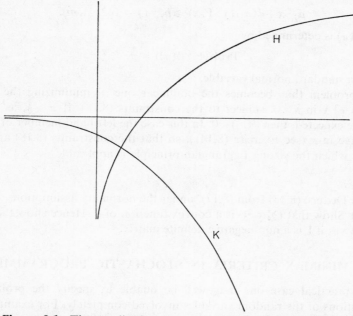

Figure 8.1 The qualitative character of the utility functions appearing in expression (8.16)

for the scalar case (i.e. respectively concave increasing and concave decreasing).

The Lagrangian aspect of the situation has now been lost. Indeed, "prices", instead of being imputed by constraints, have now to be made explicit. The optimization problem is simply a matter of unconstrained maximization.

An alternative approach sometimes adopted is to require that the constraints be satisfied with certain minimal probabilities. For example, consider the feed mix problem, in which it is required to minimize cost $c^T x$ subject to the nutritional requirements $Ax \geqslant b$. If the feed-stocks are of uncertain composition (so that A is random, but b and c are fixed and known) then one might formulate the problem as the minimization of $c^T x$ in $x \geqslant 0$ subject to

$$\text{Prob}((Ax)_j \geqslant b_j) \geqslant \pi_j \quad (j = 1, 2, ..., m). \qquad (8.17)$$

Here Prob( ) denotes the probability of the event in question, and π_j the prescribed probability that the dietary requirement on nutrient j is met.

Suppose that the column vector with elements $(a_{j1}, a_{j2}, ..., a_{jn})$ is distributed normally with mean vector μ_j and covariance matrix V_j. Then $(Ax)_j$ is distributed normally with mean $\mu_j^T x$ and variance $x^T V_j x$, and inequality (8.17) is equivalent to the inequality

$$\mu_j^T x + \psi(\pi_j)(x^T V_j x)^{\frac{1}{2}} \geqslant b_j \quad (j = 1, 2, ..., m), \tag{8.18}$$

where $\psi(\pi)$ is determined by

$$\text{Prob}(\xi \geqslant \psi(\pi)) = \pi, \tag{8.19}$$

ξ being a standard normal variable.

The problem thus becomes the nonlinear one of minimizing the linear function $c^T x$ in $x \geqslant 0$, subject to the constraints (8.18). If $\pi_j > \frac{1}{2}$, as would surely be expected, then $\psi(\pi_j) < 0$. In this case the left-hand member of (8.18) is concave in x (see Exercise (8.14)), so that the constraints (8.18) are of a type for which the strong Lagrangian principle is applicable.

Exercises

(8.13) Deduce (8.18) from (8.17), under the normality assumptions stated.

(8.14) Show that $(\sum x_k^2)^{\frac{1}{2}}$ is a convex function of x. Hence show the same for $(x^T V x)^{\frac{1}{2}}$, if V is a non-negative definite matrix.

8.5 MINIMAX CRITERIA IN STOCHASTIC PROGRAMMING

In a practical case one may well be unable to specify the probability distributions of the random variables involved completely. For example, for the stochastic allocation problem, one may have an idea of the mean values and ranges of variation of the quantities A, b and c, but scarcely more. The situation is, then, that the actual probability distribution may be any one of a class, instead of a known single distribution. Equivalently, the expectation operator E in (8.16) may be any one of a class of operators, $\mathscr{E}$.

One approach would be to protect oneself against the worst that could happen, by choosing x to maximize

$$f(x) = \min_E E[H(c^T x) + K(Ax - b)], \tag{8.20}$$

where the minimum with respect to E is taken over the class $\mathscr{E}$. That is, one considers the least possible utility that might be achieved for a given allocation x, and then chooses x so as to maximize this. Such an approach is related to the *minimax* criterion of statistical decision theory (see Section (10.5) and Exercise (8.15)).

For example, suppose it is only the resource availabilities b which are uncertain, and that we choose

$$H(c^T x) = c^T x,$$

$$K(Ax - b) = -\sum_j v_j \left(\sum_k a_{jk} x_k - b_j \right)_+$$

so that

$$f(x) = c^T x - \max_E E\left[\sum_j v_j \left(\sum_k a_{jk} x_k - b_j \right)_+ \right]. \tag{8.21}$$

If the b_j are independently distributed, and the distribution of each b_j can be separately chosen from a class of distributions $\mathscr{E}_j$, then (8.21) reduces to

$$f(x) = c^T x - \sum_j v_j \max_E E\left(\sum_k a_{jk} x_k - b_j \right)_+. \tag{8.22}$$

Suppose now that $\mathscr{E}_j$ is described in the following fashion, which represents a realistic minimum of information: that the b_j-distribution is restricted to the interval (α_j, β_j) and has mean μ_j. It is then known from Exercise (5.20) that

$$\max_E E(d_j - b_j)_+ = \begin{cases} 0 & (d_j \leqslant \alpha_j), \\ \lambda_j(d_j - \alpha_j) & (\alpha_j \leqslant d_j \leqslant \beta_j), \\ d_j - \mu_j & (d_j \geqslant \beta_j), \end{cases} \tag{8.23}$$

where

$$\lambda_j = \frac{\beta_j - \mu_j}{\beta_j - \alpha_j}. \tag{8.24}$$

If the function of d_j defined in (8.23) is denoted by $M_j(d_j)$, then it is, as seen from (8.23), a convex piecewise linear function with the representation

$$M_j(d) = \max \left\{ \begin{matrix} 0 \\ \lambda_j(d - \alpha_j) \\ d - \mu_j \end{matrix} \right\}. \tag{8.25}$$

Our criterion function (8.22) can be written

$$f(x) = c^T x - \sum_j v_j M_j \left(\sum_k a_{jk} x_k \right). \tag{8.26}$$

It is likewise concave and piecewise linear, and we can then represent it in a quasi-linear form of the type (8.9), with v taking just 3^m values.

This implies (see Section (8.3)) that the problem of maximizing $f(x)$ can be rewritten as a linear programme with 3^m constraints. It is doubtful whether such a "reduction" is often of more than formal interest, for 3^m may be a large number, and there are efficient direct methods for unconstrained

7

function maximization: see Sections (9.1)–(9.3). Nevertheless, this case is interesting as a fairly natural example of a piecewise linear criterion function, with a representation (8.9) in which ν takes finitely many values.

The material of this section is due to Žáčková (1966).

Exercise

(8.15) If the square bracketed function of (8.20) is denoted by $\phi(x)$, then note that there will be cases for which

$$\max_{x} \min_{E} E[\phi(x)] = \min_{E} \max_{x} E[\phi(x)];$$

see Section (10.4). In such cases the procedure of this section will be equivalent to an optimization with respect to x on the assumption that a given E holds, followed by the pessimist's minimization with respect to E.

8.6 THE RELATIONSHIP BETWEEN LAGRANGIAN AND FOURIER METHODS

The strong formal similarity of a maximum transform and its inverse to a Fourier transform and its inverse has already been noted in Section (7.4). In fact, Lagrangian theory generally is a particular (if extreme) case of Fourier theory, as will now be demonstrated heuristically.

Consider first the unrestricted case, where x may take any value in $\mathscr{X}$. Suppose, in fact, that x is a random variable, with a probability measure $\mu(.)$ over a suitable class of subsets of $\mathscr{X}$. Then, a more general task than that of calculating the extreme values of $f(x)$ in $\mathscr{X}$ would be to calculate the probability distribution of $f(x)$. That is, as x ranges over $\mathscr{X}$ with distribution μ, then the derived random variable $f(x)$ will range over $f(\mathscr{X})$, with a distribution which one would wish to evaluate. One key quantity in such a study would be the *characteristic function*

$$\phi(\theta) = E(e^{\theta f(x)}) = \int e^{\theta f(x)} \mu(dx) \tag{8.27}$$

(normally defined for imaginary θ, although real values will also be considered). Here E is the expectation operator. Evaluation of extreme values follows as a special case of (8.27), because, for example,

$$\max_{x} [f(x)] = \lim_{\theta \to +\infty} \frac{\log \phi(\theta)}{\theta}, \tag{8.28}$$

the limit being taken through real θ values; see Exercise (8.18).

Consider now the case when x is constrained by the m-vector equality

$$g(x) = b. \tag{8.29}$$

It is known that, granted certain assumptions, the maximum value of $f(x)$ is then given by

$$U(b) = \min_{y} [F(y) + y^T b], \tag{8.30}$$

where $F(y)$ is defined by

$$F(y) = \max_{x \in \mathcal{X}} [f(x) - y^T g(x)]. \tag{8.31}$$

In the special case $g(x) \equiv x$ this relation-pair constitutes the maximum transform and its inverse.

In the more general approach now envisaged, we would wish to calculate the probability distribution of $f(x)$ *conditional on the relation* (8.29), constraining x to a sub-set of $\mathcal{X}$. To this end we calculate the *joint characteristic function* of f and g:

$$\phi(\theta, \psi) = E[\exp(\theta f(x) + \psi^T g(x))]$$

$$= \int \exp[\theta f(x) + \psi^T g(x)] \mu(dx). \tag{8.32}$$

If $g(x)$ has a probability density, then the *conditional characteristic function* of $f(x)$ (i.e. conditional on (8.29)) is given by

$$\phi_b(\theta) = E(e^{\theta f(x)} | g(x) = b)$$

$$= \text{const} \int \phi(\theta, \psi) e^{-\psi^T b} d\psi. \tag{8.33}$$

Here "const" is a constant with respect to θ, and each element of ψ is integrated along a path in its complex plane parallel to the imaginary axis.

To reveal the closeness of this relationship to (8.30), assume θ real, and consider the new variables and functions

$$y = -\psi/\theta \tag{8.34}$$

$$e^{\theta F(y, \theta)} = \phi(\theta, -\theta y), \tag{8.35}$$

so that (8.32) and (8.33) respectively become

$$e^{\theta F(y, \theta)} = \int \exp\{\theta[f(x) - y^T g(x)]\} \mu(dx), \tag{8.36}$$

$$\phi_b(\theta) = \text{const} \, \theta^m \int \exp\{\theta[F(y, \theta) + y^T b]\} dy, \tag{8.37}$$

where the y integration path is again a complex one, each element traversing a line parallel to the imaginary axis.

Relations (8.36) and (8.37) constitute a parallel to (8.31) and (8.30) respectively, with the operation $\int \exp(\ )$ replacing the extremal operations $\max(\ )$ or $\min(\ )$. In fact, in many cases the parallelism becomes a coincidence as θ becomes large.

For, by the same argument as justifies (8.28), if θ is large and positive then the dominant contribution to the integral in (8.36) will come from x-values in the neighbourhood of values maximizing $f(x) - y^T g(x)$ and

$$F(y, \theta) = F(y) + o(1). \tag{8.38}$$

It follows likewise, from the definition of $\phi_b(\theta)$, that

$$\phi_b(\theta) = \exp[\theta U(b) + o(\theta)]. \tag{8.39}$$

Relation (8.37) will thus amount asymptotically to the "inversion" (8.30) if we can demonstrate that

$$i^{-m} \int \exp\{\theta[F(y) + y^T b]\}\, dy = \exp\left\{\theta \min_y [F(y) + y^T b] + o(\theta)\right\}. \tag{8.40}$$

This asymptotic relationship is less evident that (8.28), for example, because its justification requires an argument in the complex domain, and appeal to properties of F in this domain derived from appropriate conditions on f, g and $\mathcal{X}$.

Roughly, for θ large and positive it becomes appropriate to evaluate the integral in (8.40) by steepest descents. To do this, one shifts the y integration path laterally until it traverses a saddle-point, which it does at real y-values. At this saddle-point, whose contribution dominates the integral is locally maximal with respect to variations along the path, and so locally *minimal* with respect to variations transverse to the path (i.e. for *real* variations in y about the real saddle-point co-ordinate). Hence the minimum in the second exponent of (8.40).

In a full proof it is this point that would be the most delicate, and require special discussion: to show that the principal contribution to the integral in (8.40) comes from an appropriate saddle-point of the integrand. Some conditions on f, g and $\mathcal{X}$ are required for this to be so, and, for example, the statistical-mechanical analysis of the next section can founder unless these conditions are observed. See the comments in Exercise (8.19).

As an example, consider the case

$$f(x) = -\tfrac{1}{2} x^T C x, \tag{8.41}$$

$$g(x) = Ax, \tag{8.42}$$

where A is of full rank, C is positive definite and $\mathcal{X}$ is the whole of R^n. Then

$$F(y) = \tfrac{1}{2} y^T A C^{-1} A^T y, \tag{8.43}$$

$$U(b) = -\tfrac{1}{2} b^T (A C^{-1} A^T)^{-1} b. \tag{8.44}$$

Suppose that x has a uniform distribution in R^n, i.e. that μ is Lebesgue measure. The fact that this has infinite integral over $\mathscr{X}$ will not be of consequence unless θ is zero. It follows, by standard methods of integration, that

$$\phi(\theta, -\theta y) \overset{\text{def}}{=} e^{\theta F(y,\theta)} = \left(\frac{2\pi}{\theta}\right)^{n/2} \frac{1}{|C|^{\frac{1}{2}}} e^{\theta F(y)}, \tag{8.45}$$

so that (8.38) is confirmed. Performing the Fourier inversion, again by standard methods, we have

$$i^{-m} \int \phi(\theta, -\theta y) e^{\theta y^T b} \, dy = \left(\frac{2\pi}{\theta}\right)^{(m+n)/2} \frac{e^{\theta U(b)}}{|C|^{\frac{1}{2}}|AC^{-1}A^T|^{\frac{1}{2}}}$$

$$= \exp\left[\theta U(b) + o(\theta)\right], \tag{8.46}$$

so that asymptotic equivalence of (8.37) to (8.30) is again confirmed.

The analogue (and, under suitable conditions, asymptotic identity) of the Fourier and Lagrangian theory persists in the case of inequality constraints: see Exercise (8.17).

No good reason has been given for considering large θ, save that this is the extreme case in which the maximum of $f(x)$ predominates (see (8.28)) and one then can expect Fourier theory to yield Lagrangian results. In the next section we shall consider a physical example for which the motivation is more natural, and the significance of these results may be clearer.

Exercises

(8.16) Confirm equations (8.45) and (8.46).

(8.17) Suppose that the constraint (8.29) is replaced by $g(x) \leqslant b$, so that the minimization in (8.30) is to be taken only over non-negative y. Then formula (8.37) will be replaced by

$$E(e^{\theta f(x)} \,|\, g(x) \leqslant b) = \lim_{\beta \to -\infty} \text{const} \int_\beta^b \phi_\xi(\theta) \, d\xi$$

$$= \lim_{\beta \to -\infty} \text{const} \int e^{\theta F(y,\theta)} \prod_j \left[(e^{\theta b_j y_j} - e^{\theta \beta_j y_j}) y_j^{-1} \, dy_j\right].$$

When inverting this conditioned characteristic function one must choose a y integration path which observes the restriction $\text{Re}(y) \geqslant 0$ or there is trouble with the terms $e^{\theta \beta_j y_j}$ as $\beta \to -\infty$. This correlates with the restriction $y \geqslant 0$ mentioned above.

Demonstrate the corresponding result for the case $b - Ax \in \mathscr{C}$, with $\mathscr{C}$ a convex cone.

(8.18) Relation (8.28) does not hold wholly without regularity conditions on f and μ. Let $\bar{f}$ denote the supremum of f in $\mathscr{X}$, and suppose that the set of x satisfying $f(x) \geqslant \bar{f} - \varepsilon$ has strictly positive probability for any $\varepsilon > 0$. Then show that the limit in equation (8.28) equals $\bar{f}$.

8.7   CHEMICAL EQUILIBRIUM: AN APPLICATION OF LAGRANGE/FOURIER THEORY

The problem is: given m chemical elements in prescribed abundances, from which n compounds can be formed, what will be the distribution of the amounts of the various compounds formed when statistical equilibrium at a given temperature has been reached?

Suppose that a molecule of compound k contains a_{jk} molecules of element j. Consider now a volume θ, in which there are N_k molecules of compound k. If

$$x_k = N_k / \theta \tag{8.47}$$

then x_k is the density of compound k, in number of molecules per unit volume. The N_k, or x_k, are constrained by

$$\sum_k a_{jk} N_k = \theta b_j \tag{8.48}$$

or

$$\sum_k a_{jk} x_k = b_j \quad (j = 1, 2, ..., m), \tag{8.49}$$

where b_j is the prescribed density of element j. Prescription of temperature is equivalent to prescription of total energy in the volume θ; if energy is regarded as just another "element", of prescribed abundance, then this constraint can be assumed incorporated in the set (8.49).

The argument starts from an assertion basic in statistical mechanics, that the probability of $N_1, N_2, ..., N_n$ molecules of the respective compounds in the volume θ is proportional to

$$\text{Prob}(N) = \text{const} \prod_k \theta^{N_k} / N_k! \tag{8.50}$$

if the N_k are non-negative integers satisfying (8.48); otherwise, the probability is zero. The constant is such as to normalize this conditional distribution.

Consider first, then, the determination of the most likely value of the N_k, conditional on the constraints (8.48). If θ is large, then the maximizing N_k will also be large (of order θ), so that $N!$ can be approximated by the crude form of Stirling's formula

$$\log N! = N \log N - N + o(N). \tag{8.51}$$

The aim is then to maximize

$$\log [\text{Prob}(N)] \sim \text{const} + \sum_k N_k (1 + \log (\theta / N_k)) \tag{8.52}$$

subject to (8.48), or, equivalently in the limit, to maximize

$$f(x) = \sum_k x_k(1 - \log x_k) \tag{8.53}$$

in $x \geq 0$, subject to (8.49).

Since f is concave and the constraints are linear, one can locate the maximizing x-value by maximizing the Lagrangian form

$$L(x, y) = \sum_k x_k(1 - \log x_k) + \sum_j y_j\left(b_j - \sum_k a_{jk} x_k\right), \tag{8.54}$$

obtaining

$$\bar{x}_k = \exp\left(-\sum_j a_{jk} y_j\right) = \prod_j z_j^{a_{jk}}, \tag{8.55}$$

say, and

$$\max_x L(x, y) = \sum_k \exp\left(-\sum_j a_{jk} y_j\right) + \sum_j b_j y_j. \tag{8.56}$$

If the maximization of (8.53) subject to (8.49) is taken as the primal problem, the dual is then to minimize expression (8.56) with respect to real y_j, or to minimize

$$Q(z) = \sum_k \prod_j z_j^{a_{jk}} - \sum_j b_j \log z_j \tag{8.57}$$

with respect to real positive z_j. The primal and dual are clearly closely related to the dual and primal respectively of the geometric programming problem treated in Section (6.4), where the validity of the dualization is justified in detail.

Now, the primal problem (the constrained maximization of $f(x)$) emerged as the limit case ($\theta \to \infty$) of a distributional problem (the conditioned distribution of the N_k). It is then natural to ask what interpretation can be given to its dual (minimization of (8.56)) in the same distributional context.

If the N_k followed the unconditioned distribution const $\Pi \theta^{N_k}/N_k!$, then the joint probability generating function of the numbers of compound molecules N_k and element atoms

$$M_j = \sum_k a_{jk} N_k \tag{8.58}$$

in the volume θ would be proportional to

$$\Pi(w, z) = \sum_{N \geq 0} \left(\prod_k (\theta w_k)^{N_k}/N_k!\right)\left(\prod_j z_j^{M_j}\right)$$

$$= \exp\left[\theta \sum_k w_k \prod_j z_j^{a_{jk}}\right]. \tag{8.59}$$

But the distribution is constrained by (8.49), i.e. by $M_j = \theta b_j$, so the conditional probability generating function of the N_k will be proportional to the coefficient of $\prod_j z_j^{\theta b_j}$ in the expansion of (8.59). Evaluating the coefficient by a multiple contour integral, one sees that this conditional p.g.f. is

$$\Pi_b(w) = \text{const} \int \exp\left[\theta \sum_k w_k \prod_j z_j^{a_{jk}} \right] \prod_j z_j^{-1-\theta b_j}\, dz_j, \tag{8.60}$$

where the z_j integration paths each form a small circle around the origin in the complex z_j-plane. (These circles correspond to the y_j paths of Section (8.6) parallel to the imaginary axis.)

So, for example, the conditioned mean value of N_k will be given by

$$E(N_k \,|\, M = \theta b) = \left[\frac{\partial \pi_b(w)}{\partial w_k} \right]_{w=1}$$

$$= \frac{\theta \int e^{\theta Q(z)} \prod\limits_j z_j^{a_{jk}-1}\, dz_j}{\int e^{\theta Q(z)} \prod\limits_j z_j^{-1}\, dz_j}, \tag{8.61}$$

where $Q(z)$ is just expression (8.57). For θ large the integrals in (8.61) will be evaluated by steepest descents. This means that, if possible, the circular integration paths will be varied until they pass through a saddle-point, which they will do at a positive real value. At this point $Q(z)$ will be maximal along the integration path, or minimal in a direction transverse to it—for real positive z. Expression (8.61) will then be identical with (8.55), the z_j being characterized as the values minimizing expression (8.57).

Thus, by considering the distributional problem in this generating function (i.e. Fourier) form, one is led in the limit of large θ to the dual form of the primal version previously encountered. The attacks on the problem via primal or dual correspond to familiar alternative approaches in statistical mechanics: the calculation of the most probable values (of the N_k or x_k), or the calculation of conditional mean values.

If there are infinitely many types of compound, so that $Q(z)$ is defined by an infinite series in (8.57), then a phenomenon is possible which cannot occur in the case of finite n. As the densities b are increased, the value of z minimizing $Q(z)$ may tend to a singular point of $Q(z)$. When this singular point is reached, the equilibrium statistics change character entirely, and there is a *change of state* (usually; compounds with very large molecules suddenly becoming dominant). This transition becomes sharper the larger θ, and in the limit there is a well-defined critical point, along some locus of critical densities b_j.

Exercises

(8.19) The constrained maximization problem of Exercise (2.5) is a special case of that treated in this section, with the constraints on total particle numbers and energy being equivalent to specification of the abundances of two "elements". Nevertheless, verify for this particular case the emergence of the primal and dual problems as limit versions of the two possible attacks on the distributional problem in the limit of large volume ($N, \mathscr{E} \to \infty$ in constant proportion).

If the energy levels ε_k are such that the two constraints $\sum x_k = N$, $\sum x_k \varepsilon_k = \mathscr{E}$ have few solutions in integral x_k, then these asymptotic evaluations may be invalid: the sum $\sum \exp(-y_2 \varepsilon_k)$ has then such a character that the steepest descent evaluation in the analogue of (8.61) fails. What one does in such a case is to let the average energy $\mathscr{E}/N$ adopt values in a narrow band, when the formal calculations become valid. From one point of view, the constraints then have sufficiently many solutions; from another, a damping factor appears inside the integral, which suppresses contributions from all but the real saddle-point.

(8.20) The quantity being maximized in Exercise (6.8) is evidently an approximation (large x) to the logarithm of

$$\left(\prod_j S_j! \right) \left(\prod_j c_k^{x_k} \middle/ x_k! \right).$$

Interpret the constrained maximization of this quantity as the determination of a most probable configuration under conservation constraints, and interpret the dual.

CHAPTER 9

Numerical Methods

To many authors, optimization means numerical optimization; an optimization problem is seen as so much computer-grist, and "solution" means a grinding in the computer mill, after a preliminary reduction to lumps of convenient size. It is rather in reaction to this view that we have stressed analytic methods, and followed the analytic road to the end, as far as could be seen, in each case.

However, for every problem one must resort to numerical methods at one stage or another. Moreover, having arrived at this point, one finds that the problems of numerical analysis are themselves interesting, and sometimes deep.

The only numerical methods we have described hitherto are those designed for linear programming problems: the simplex method and variants of it (Sections (4.5)–(4.8)). What is now required is a numerical maximum-seeking method for a moderately general function $f(x)$, for both unconstrained and constrained cases.

Virtually all methods for the unconstrained case are so-called "gradient" methods (see Section (9.2)). Of these the elegant Fletcher–Powell algorithm (Section (9.3)), which bids fair to become the definitive standard, is described in detail. All these methods work by reducing the problem to a succession of one-dimensional problems (i.e. of maximizations with respect to a scalar argument), so the one-dimensional case is treated separately in Section (9.1).

A host of methods has been proposed for dealing with numerical maximization under constraints. Of these only one class is considered: the sequential unconstrained methods. In these a penalty $R(x)$ is attached to those values of x which do not satisfy the constraints, and then $f(x) - R(x)$ is maximized unconstrainedly. A sequence of problems is considered in which the penalty grows progressively stiffer: under appropriate conditions the sequence of solutions then converges to the solution of the constrained problem.

192

These sequential methods seem to work very well in practice; they are general in application, and provide a natural step from the unconstrained case. Moreover, they are closely related to Lagrangian methods, as will be seen. There are indeed a great many other methods, some having virtues for particular cases, but it seems likely that the sequential methods will establish themselves in the first rank of the general algorithms.

The one special case demanding attention is that of linear programming: this has been treated in Chapter 4. Other special areas with their own numerical methods, which have generated a large literature, are quadratic programming, integer programming, scheduling, etc. These are not covered here; the interested reader is referred to texts such as those by Beale (1968) or Künzi, Tzschach and Zehnder (1968). For the more recent sequential methods, see Fiacco and McCormick (1968).

9.1 MAXIMUM-SEEKING METHODS IN ONE DIMENSION

Consider a function $f(x)$ of scalar argument x on an interval AB, and suppose that one wishes to locate the maximizing point $\bar{x}$ in this interval. In order to be able to draw any conclusions at all from a finite number of evaluations of $f(x)$, one must restrict the behaviour of the function: $f(x)$ will be assumed unimodal in AB, and not constant on any sub-interval of AB. So, the graph of $f(x)$ will either have the general character depicted in Figure (9.1) or be strictly monotonic.

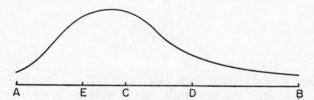

Figure 9.1 Location of the maximum of a unimodal function by sequential methods

If one is prepared to evaluate $f(x)$ at r points, then one's first impulse is to space these points so that the interval AB is divided into $r+1$ equal points. Since $\bar{x}$ must lie in one of the sub-intervals adjacent to a point at which the evaluation of f is largest, $\bar{x}$ is located within an interval whose length is in general a fraction $2/(r+1)$ of AB. As will be seen below, this is very inefficient. A sequential method, in which the choice of the next point at which f is calculated depends on the result of past evaluations, can do very much better.

Suppose that one begins with evaluations at two points C and D of AB. If

$$f(x_C) \geqslant f(x_D), \qquad (9.1)$$

in an obvious notation, then $\bar{x}$ must lie in the interval AD. Thus, with the two evaluations, $\bar{x}$ can certainly be located in one of the intervals AD or CB, and it would seem reasonable to choose these equal in length.

Suppose (9.1) holds, so that attention can be restricted to the interval AD. This already contains one evaluation, at C, and another will be made, at a point E, say. E will be chosen so that $AC = ED$, as before.

Now, the procedure would have a simple invariant character if the points were chosen in such a way that the current sub-interval of interest were divided in the same ratios at all stages. For this C (and so automatically D) should be chosen so that C divides DA in the same ratio as it divides AB. That is, so that

$$\frac{AC}{AD} = \frac{AD}{AB}. \qquad (9.2)$$

If

$$\rho = \frac{AD}{AB}, \qquad (9.3)$$

we have then

$$\frac{1-\rho}{\rho} = \rho$$

or

$$\rho = \frac{\sqrt{5}-1}{2} \sim 0.618, \qquad (9.4)$$

the so-called "golden section" (see Exercise (9.2)).

At each stage after the first, then, one is working in an interval already divided by an evaluation in the ratio $\rho : 1$, and one makes the new evaluation at the symmetrically placed point in the interval, dividing it in the ratio $1 : \rho$. This method is sometimes termed "golden section search". At each stage after the first the new evaluation reduces the length of the interval of interest by a factor of ρ. With r observations $(r > 1)$ $\bar{x}$ can thus be located in a sub-interval whose length is a fraction ρ^{r-1} of AB: a tremendous improvement on the fraction $2/(r+1)$ achieved by the non-sequential method.

It can be shown (see Exercises (9.4)–(9.6)) that this method is very near the optimal. The optimal method (in the minimax sense: that the length of the interval within which $\bar{x}$ can be guaranteed to lie after r evaluations is minimal) is the so-called method of *Fibonacci search* derived by Kiefer (1953). With this method the initial interval is divided in the ratios $F_{r-1} : F_{r-2}$ and $F_{r-2} : F_{r-1}$ the next sub-interval is divided in the ratios $F_{r-2} : F_{r-3}$ and inversely, and so on. Here F_r is the rth Fibonacci number, determined by $F_0 = F_1 = 1$ and

$$F_r = F_{r-1} + F_{r-2} \quad (r > 1). \tag{9.5}$$

Solving this difference equation, one finds that

$$F_r = \frac{\rho^{-r} + (-\rho)^{r+2}}{1 + \rho^2}, \tag{9.6}$$

where ρ is given by (9.4). Thus, for large r

$$F_r \sim \frac{\rho^{-r}}{1 + \rho^2} \tag{9.7}$$

and F_{r-1}/F_r is asymptotic to ρ. The optimal strategy thus tends to the fixed or golden section strategy with increasing r. The reduction factors achieved by the two methods, with r observations, are $F_r^{-1} \sim (1 + \rho^2) \rho^r$ and ρ^{r-1} respectively. The size of the final interval for golden section search thus exceeds that deduced by the optimal method by a factor of

$$\frac{\rho^{r-1}}{(1 + \rho^2) \rho^r} = \frac{1}{\rho(1 + \rho^2)} \sim 1.171 \tag{9.8}$$

Thus, the fixed strategy of dividing intervals in the ratio $\rho^{\pm 1} : 1$ produces a final uncertainty of the same order as does the optimal strategy, but about 17% larger. On the other hand, its invariant nature makes it simpler to apply.

Exercises

(9.1) Show that the reduction factor of $2/(r+1)$ is the best that can be guaranteed with a non-sequential allocation of evaluation points.

(9.2) The Greeks considered the rectangle of most harmonious proportions to be that whose proportions were conserved when the square erected on the shorter side was removed. Note that the shorter side of such a rectangle is a fraction ρ of the longer side, with ρ given by (9.4), defining the "golden section".

(9.3) What was the necessity for the two conditions on $f(x)$?

(9.4) Let $L_s(a, \xi)$ denote the minimal length of the interval within which $\bar{x}$ can be guaranteed to lie, if one knows that it lies in the interval $(0 \leqslant x \leqslant a)$, one has an evaluation of $f(\xi)$ (where ξ lies in $(0, a)$) and s further evaluations are permitted. Show that

$$L_s(a, \xi) = \min \left\{ \begin{array}{l} \min_{\eta \leqslant \xi} \max [L_{s-1}(\xi, \eta), L_{s-1}(a - \eta, a - \xi)] \\ \min_{\eta \geqslant \xi} \max [L_{s-1}(\eta, \xi), L_{s-1}(a - \xi, a - \eta)] \end{array} \right\},$$

where the minimizing η is the optimal choice of the next evaluation point.

(9.5) Show that the recursion of Exercise (9.4) has solution

$$L_s(a, \xi) = \begin{cases} \max\left(\dfrac{\xi}{F_s}, \dfrac{a-\xi}{F_{s-1}}\right) & (\xi \geqslant a/2), \\[3mm] \max\left(\dfrac{a-\xi}{F_s}, \dfrac{\xi}{F_{s-1}}\right) & (\xi \leqslant a/2), \end{cases}$$

where F_s is the sth Fibonacci number, as defined in the text.

(9.6) Show that $L_s(a, \xi)$ as determined in Exercise (9.4) is minimal with respect to ξ if either ξ or $a-\xi$ equals $(F_s/F_{s+1})a$. Hence show that the Fibonacci search method defined in the text is optimal in the minimax sense.

Note an implication of Fibonacci search: that the last two evaluations are both placed at the mid-point of the remaining interval. Essentially, it is supposed that if two observations are permitted in an interval of length a, then observations at the points $\frac{1}{2}a+$ and $\frac{1}{2}a-$ will locate the maximum within an interval of $\frac{1}{2}a$. That is, it is supposed that the calculation has infinite resolution, and values of the function at indefinitely close values of the argument can be distinguished. However, the case of limited resolution can be coped with, by the same methods. It is left to the reader to adapt this and the previous two exercises to the case when the x-values at which f is evaluated must be separated by at least an amount δ.

9.2 GRADIENT METHODS IN SEVERAL DIMENSIONS

Consider the problem of maximizing a function $f(x)$ of several variables, without constraints.

The general technique adopted is to take a fixed direction of search, s, and to then consider the one-dimensional problem of maximizing $f(x)$ in this direction. That is, if the starting point is x_0 and the initial search direction is s_0, one considers $f(x_0 + \theta s_0)$ for variable scalar θ, and maximizes this with respect to θ by a one-dimensional maximum-seeking method of the type described in the previous section. This leads to a new trial point x_1, from which one searches again in a new direction s_1. One hopes that the sequence of trial solutions $\{x_i\}$ thus generated converges to the global maximum $\bar{x}$ of $f(x)$, although this can be guaranteed only under rather special circumstances (e.g. concave f). Note that x_i, s_i denote particular values of the vectors x, s, not elements of them, in this section and the next.

The natural direction of search to take is that in which $f(x)$ increases fastest at the current point: the so-called *method of steepest ascents*. This will be the direction of the gradient at x, so that s_i will be proportional to

$$d_i = (f_x^{\mathrm{T}})_{x=x_i}. \tag{9.9}$$

However, while this method can sometimes work very well, it can also work very badly. Suppose that the graph of $f(x)$ has a long, sharp, curved ridge, as illustrated in the contour map of Figure (9.2). The steepest ascent

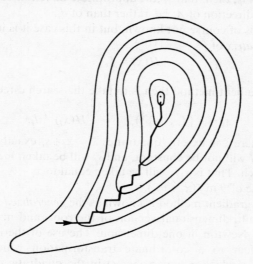

Figure 9.2 In cases where the graph of the function being maximized presents a narrow curved ridge, the method of steepest ascents leads to an ascent path consisting of many short segments

path can follow this ridge only by making a large number of short traverses, in a zig-zag pattern, which gives slow convergence to the hill-top.

Ideally, one would wish for a standardizing transformation of co-ordinates which would even out possible variations in curvature of the $f(x)$ surface, and so eliminate the difficult ridge phenomenon. A valuable pointer in this direction is provided by considering the special case where f is a quadratic form:

$$f(x) = -\tfrac{1}{2} x^T A x + b^T x + c. \tag{9.10}$$

Here A must be non-negative definite if f is to have a finite maximum: positive definite if the maximum is moreover to be unique.

The gradient at x_i is

$$d_i = b - A x_i. \tag{9.11}$$

The co-ordinate of the maximum is

$$\bar{x} = A^{-1} b, \tag{9.12}$$

so that

$$\bar{x} = x_i + A^{-1} d_i. \tag{9.13}$$

The intimation is, then, that s_i, the appropriate direction of search from x_i, should be in the direction of $A^{-1} d_i$ rather than of d_i.

The matrix A is of course not known, but in this case it is just the negative of the *Hessian matrix* of f

$$H(x) = f_{xx} \tag{9.14}$$

for all x.

A modified gradient method then is to take the search direction from x_i as being

$$s_i = -(f_{xx}{}^{-1} f_x{}^T)_{x=x_i} = -H(x_i)^{-1} d_i. \tag{9.15}$$

For the case where f is exactly quadratic, $\bar{x} = x_i + s_i$ exactly. For the non-quadratic case H will not be constant, and s_i will be taken merely as the next direction of search. That is x_{i+1} will be taken equal to $x_i + \theta_i s_i$ where θ_i is the maximizing value of θ in $f(x_i + \theta s_i)$.

This modified gradient method is known as the *generalized Newton method*, since it is the multi-dimensional version of the root- and maximum-seeking method used by Newton in one dimension. The use of the inverse Hessian, H^{-1}, is equivalent to a co-ordinate transformation which standardizes contour lines to circles (or hyperspheres) in the quadratic case (9.10). The hope is that, in more general cases, it will still tend to remove the ridge singularities so embarrassing to the simple steepest ascent method. In this it is usually effective.

However, the recomputation of the Hessian and its inverse at every step is a heavy task. For this reason, considerable attention has been given to methods which begin with an approximate guessed value of H^{-1}, and use evaluations of f to improve and update this approximation, without matrix inversion. The most successful of these is the Fletcher–Powell method, described in the next section.

The following preparatory result is useful for the case when knowledge of the Hessian is built up step by step. It concerns the so-called *method of conjugate gradients*.

Theorem (9.1)

Let x be an n-vector and $f(x)$ the negative-definite quadratic form (9.10). Suppose that the test directions s_i are chosen mutually conjugate in the sense that

$$s_i{}^T A s_j = 0 \quad (i \neq j). \tag{9.16}$$

Then (i) *The sequence $\{x_i\}$ converges to the maximizing value $\bar{x}$ in at most n steps.*

(ii) *Conjugacy is equivalent to orthogonality of s_i to*

$$d_j - d_{j-1} \quad (j < i; \; i = 1, 2, \ldots, n-1).$$

(iii) *If $s_0, s_1, \ldots, s_{i-1}$ are mutually conjugate, then $x_i - \bar{x}$ is conjugate to s_j and d_i is orthogonal to s_j ($j < i$).*

For, we have

$$x_i = x_0 + \sum_{j=0}^{i-1} \theta_j s_j, \tag{9.17}$$

and, by virtue of (9.10), (9.16) and (9.17),

$$f(x_i) = f(x_0) + \sum_{0}^{i-1} \theta_j (b - Ax_0)^{\mathrm{T}} s_j - \frac{1}{2} \sum_{0}^{i-1} \theta_j^2 (s_j^{\mathrm{T}} As_j), \tag{9.18}$$

whatever the values of the θs. Regarded as a function of the θs, this has the separable form

$$f(x_i) = \text{const} + \sum_{0}^{i-1} \phi_j(\theta_j) \tag{9.19}$$

so that if θ_{i-1} is chosen to maximize $f(x_{i-1} + \theta_{i-1} s_{i-1}) = f(x_i)$, as is the determining rule, it maximizes $\phi_{i-1}(\theta_{i-1})$. Equivalently, the $\theta_0, \theta_1, \ldots, \theta_{n-1}$ thus determined maximize $f(x_n) = \text{const} + \sum_{0}^{n-1} \phi_j(\theta_j)$ jointly. But this is equivalent to a free maximization of $f(x_n)$ with respect to x_n, because relation (9.17) for $i = n$ constitutes a non-singular linear relationship between x_n and $\theta_0, \theta_1, \ldots, \theta_{n-1}$. That is, sequential maximization with respect to $\theta_0, \theta_1, \ldots, \theta_{n-1}$ achieves the same result as a free maximization with respect to x_n, which is just assertion (i) of the theorem.

Of course, if A is not known, then the s_i cannot be chosen in advance to satisfy (9.16). Assertion (ii) covers this point. Define the increments in co-ordinate

$$\sigma_i = x_{i+1} - x_i = \theta_i s_i \tag{9.20}$$

and in gradient

$$y_i = d_i - d_{i+1}, \tag{9.21}$$

so that mutual conjugacy of the σ_i is implied by that of the s_i. In the quadratic case

$$y_i = A\sigma_i, \tag{9.22}$$

by (9.11), so that, by choosing s_{i+1} orthogonal to $y_1, y_2, \ldots, y_i$ one ensures that s_{i+1} is conjugate to $s_1, s_2, \ldots, s_i$. That is, mutual conjugacy is assured if the next increment in position is always chosen orthogonal to past increments in gradient, as asserted in (ii).

With optimal choice of θs we have

$$\bar{x} - x_i = x_n - x_i = \sum_i^{n-1} \theta_j s_j, \qquad (9.23)$$

and also, by (9.13),

$$d_i = A(\bar{x} - x_i) = \sum_i^{n-1} \theta_j A s_j. \qquad (9.24)$$

Assertion (iii) follows from (9.23), (9.24) and the mutual conjugacy of the s_i. Only the mutual conjugacy of $s_0, s_1, \ldots, s_{i-1}$ is invoked in assertion (iii), because, at this stage in the calculation, $s_i, \ldots, s_n$ may not have been determined. The point is that they can be determined, consistent with (9.16), and that representations (9.23) and (9.24) will then hold.

Of course, it is not imagined that f will be quadratic or even near-quadratic (except locally) in practical cases. However, it is likely that methods which are efficient for the quadratic case will be good generally, and experience bears this out.

By the same token, one cannot in general be sure that the limit to which the trial sequence $\{x_i\}$ tends is the co-ordinate of the global maximum of f. It will certainly be the co-ordinate of a local maximum, but its global character cannot be guaranteed. The usual test applied is to run the maximum-seeking computation from a number of widely separated starting points x_0, and to check that these computations all lead to the same solution point.

9.3   THE FLETCHER–POWELL TECHNIQUE

Starting from an initial approximate guess, G_0, the Fletcher–Powell technique generates a sequence of approximations $\{G_i\}$ to the negative inverse Hessian

$$G = -H^{-1} \qquad (9.25)$$

such that the directions

$$s_i = G_i d_i \qquad (9.26)$$

are automatically mutually conjugate, and $G_n = A^{-1}$ if f is given by (9.10). Because of mutual conjugacy, $x_n = \bar{x}$.

If σ_i is the x-increment defined in (9.20), then the σ_i are mutually conjugate,

$$\sigma_i^T A \sigma_j = 0 \quad (i \neq j), \qquad (9.27)$$

if the s_i are. Relation (9.27) can be written

$$\sigma^T A \sigma = \Lambda, \qquad (9.28)$$

if σ is the matrix with columns $\sigma_0, \sigma_1, ..., \sigma_{n-1}$, and Λ the diagonal matrix with entries $\sigma_i^T A \sigma_i$.

(9.28) can be alternatively written as

$$G = \sigma \Lambda^{-1} \sigma^T = \sum_0^{n-1} \lambda_i^{-1} \sigma_i \sigma_i^T \tag{9.29}$$

where $G = A^{-1}$, the constant value of the negative inverse Hessian in the quadratic case (9.10). The conjugacy assumption (9.27) is thus equivalent to the representation (9.29) of G. Since by (9.28) and (9.22)

$$\lambda_i = \sigma_i^T A \sigma_i = \sigma_i^T y_i, \tag{9.30}$$

this representation can be rewritten as

$$G = \sum_i \frac{\sigma_i \sigma_i^T}{\sigma_i^T y_i}. \tag{9.31}$$

Theorem (9.2)

Let the sequence of test directions $\{s_i\}$ be generated by (9.26), and the sequence $\{G_i\}$ by

$$G_{i+1} = G_i - \frac{(G_i y_i)(G_i y_i)^T}{y_i^T G_i y_i} + \frac{\sigma_i \sigma_i^T}{\sigma_i^T y_i} \quad (i = 0, 1, 2, ...). \tag{9.32}$$

Suppose that f has the negative definite quadratic form (9.10). Then, $\sigma_0, \sigma_1, ..., \sigma_{n-1}$ are mutually conjugate, and G_i has the form

$$G_i = F_i + \sum_{j<i} \frac{\sigma_j \sigma_j^T}{\sigma_j^T y_j} \quad (i \leqslant n), \tag{9.33}$$

where F_i is a symmetric matrix satisfying

$$F_i A \sigma_j = 0 \quad (j < i). \tag{9.34}$$

Thus $x_n = \bar{x}$ and $G_n = G = A^{-1}$.

The recursion (9.32) obviously corresponds to an attempt to bring G_i nearer the form (9.31): successful, in view of (9.33). The last assertions of the theorem follow from the earlier ones. For, if the vectors $\sigma_0, \sigma_1, ..., \sigma_{n-1}$ are mutually conjugate, then $x_n = \bar{x}$, by Theorem (9.1). They will then also be linearly independent, so that (9.34) will imply that $F_n = 0$, and (9.31) and (9.33) will then imply that $G_n = G$.

The theorem will be proved inductively, by assuming the validity of (9.33) and the mutual conjugacy of the σ_j $(j < i)$ for a given value of i. These assertions are trivially true for $i = 0$.

Note first that the s_i of (9.26) (and so σ_i) is conjugate to $\sigma_i (j < i)$, for

$$\sigma_j^T A s_i = \sigma_j^T A G_i d_i = \sigma_j^T d_i \quad (j < i), \tag{9.35}$$

from (9.26) and (9.33). But, by assertion (iii) of Theorem (9.1), expression (9.35) is zero, whence the conjugacy of σ_i to earlier σ_j follows. Note that this implies that

$$G_i y_i = F_i y_i. \tag{9.36}$$

Consider now the updating of G_i by

$$G_{i+1} = G_i - \gamma\gamma^{\mathrm{T}} + \frac{\sigma_i \sigma_i^{\mathrm{T}}}{\sigma_i^{\mathrm{T}} y_i}, \tag{9.37}$$

where γ is a vector so chosen that $G_i - \gamma\gamma^{\mathrm{T}}$ contains no term in $\sigma_i \sigma_i^{\mathrm{T}}$, in the sense that

$$(G_i - \gamma\gamma^{\mathrm{T}}) A\sigma_i = 0. \tag{9.38}$$

Since A is not known, we shall rather replace condition (9.38) by

$$(G_i - \gamma\gamma^{\mathrm{T}}) y_i = 0. \tag{9.39}$$

This determines γ as

$$\gamma = \pm \frac{G_i y_i}{(y_i^{\mathrm{T}} G_i y_i)^{\frac{1}{2}}}, \tag{9.40}$$

consistently with (9.32). Now, recursion (9.32) applied to (9.33) implies that F_{i+1} must be of the form

$$F_{i+1} = F_i - \frac{(F_i y_i)(F_i y_i)^{\mathrm{T}}}{y_i^{\mathrm{T}} F_i y_i} \tag{9.41}$$

(see (9.36)). It is readily verified that, with this determination,

$$F_{i+1} y_j = F_{i+1} A\sigma_j = 0 \quad (j < i+1), \tag{9.42}$$

so that the induction on the representation (9.33) of G_i and the mutual conjugacy of $\sigma_0, \sigma_1, \ldots, \sigma_{i-1}$ is complete.

Of course, all these assertions concern the quadratic case (9.10), for which it is guaranteed that $x_n = \bar{x}$ and $s_n = d_n = 0$. For the general case this will not be so, and iterations (9.26) and (9.32) must be continued until $d = f_x^{\mathrm{T}}$ is suitably small. However, the method is found to be rapid and effective in practice. It is attractive in that it generates the sequence of approximations to the inverse Hessian $\{G_i\}$ without matrix inversions or evaluation of second derivatives; this sequence is guaranteed to converge to the true value in n steps in the quadratic case, and the test directions generated by (9.26) are automatically mutually conjugate.

For explicitness, the numerical procedure is summarized.

(i) One starts with trial values x_0 and G_0 of the solution point and the negative inverse Hessian: at the completion of the ith iteration one will have values x_i, G_i and $d_i = (f_x^{\mathrm{T}})_{x=x_i}$.

(ii) Calculate s_i from (9.26).

(iii) Calculate $x_{i+1} = x_i + \theta_i s_i$ by maximizing $f(x_i + \theta_i s_i)$ with respect to the scalar θ_i.

(iv) Calculate d_{i+1}, also $\sigma_i = x_{i+1} - x_i$ and $y_i = d_i - d_{i+1}$.

(v) Calculate G_{i+1} from (9.32).

The steps (ii)–(v) are then repeated at the values pertaining to $i+1$.

9.4 CONSTRAINED MAXIMIZATION; PENALTY FUNCTIONS AND LAGRANGIAN METHODS

Consider the maximization of $f(x)$ under the constraint that x must be chosen in a *feasible set* $\mathcal{F}$, smaller than the basic set $\mathcal{X}$. That is, $\mathcal{F}$ is a proper subset of $\mathcal{X}$. One natural approach is to reformulate this as the free maximization of $f(x) - R(x)$ over $\mathcal{X}$, where R is a *penalty function* having a behaviour at least approximating the prescription

$$R(x) = \begin{cases} 0 & (x \in \mathcal{F}), \\ +\infty & (x \in \overline{\mathcal{F}}), \end{cases} \tag{9.43}$$

$\overline{\mathcal{F}}$ being the complement of $\mathcal{F}$ in $\mathcal{X}$.

In fact, it is computationally impossible to handle functions of this nature, so the procedure is rather to consider a family of penalty functions $R(x, \lambda)$ which are well behaved for finite λ, but which tend to limit (9.43) as $\lambda \to +\infty$. One chooses a sequence $\{\lambda^{(r)}\}$ increasing to infinity, and successively maximizes the forms $\phi(x, \lambda^{(r)}) = f(x) - R(x, \lambda^{(r)})$ freely in $\mathcal{X}$. The maximizing value $x^{(r)}$ is used as an initial trial solution for the maximization of the next form, $\phi(x, \lambda^{(r+1)})$. The hope is that, as r increases, $x^{(r)}$ will approximate arbitrarily closely to a solution of the original problem: a point proved under certain assumptions in Theorem (9.3).

For example, suppose the restrictions take the form of the familiar m-vector inequality

$$g(x) \leqslant b. \tag{9.44}$$

We might then take

$$R(x, \lambda) = \sum_{j=1}^{m} \exp \left[\lambda_j (g_j(x) - b_j) \right], \tag{9.45}$$

where λ is an m-vector with elements λ_j. As $\lambda \to \infty$ (i.e. $\lambda_j \to +\infty$ for all j) then $R(x, \lambda)$ tends to the limit (9.43), unless (9.44) holds with equality for s elements, when $R \to s$. These exceptional cases correspond to points x on the boundary of $\mathcal{F}$, and cause only minor complications.

Another obvious penalty function for the case (9.44) would be

$$R(x, \lambda) = \sum_{j=1}^{m} \lambda_j (g_j(x) - b_j)_+, \tag{9.46}$$

which certainly tends to the $R(x)$ of (9.43) as the λ_j tend to $+\infty$.

The penalty function approach (a numerical technique) and the Lagrangian approach (an analytic technique) can be united in the concept of a *generalized Lagrangian function* (Roode, 1968). By this is meant a function $\phi(x, y)$ with properties which ensure that the problem of maximizing $f(x)$ subject to constraints can be replaced by that of maximizing $\phi(x, y)$ freely, for some y. The classic Lagrangian $L(x, y)$ of (3.6) is a special case of such a function. There are cases for which modified Lagrangian approach is effective where the classic one fails, although the classic Lagrangian, generating as it does the notions of "price" and of randomized solutions, does seem to have a special status.

General theorems exist (see Roode, 1968) to the effect that, under certain conditions, $\phi(x, y)$ is both maximal in x and minimal in y at the solution point (i.e. that the solution point is a *saddle-point* of ϕ; see Section (10.4)). This leads to a duality theory generalizing that of Section (3.7). It also leads to a sequential algorithm of the primal–dual type: two sequences $\{x^{(r)}\}, \{y^{(r)}\}$ are determined; $x^{(r)}$ being the value of x maximizing $\phi(x, y^{(r)})$ over $\mathscr{X}$, and $y^{(r+1)}$ being the value of y minimizing $\phi(x^{(r)}, y)$ over some appropriate set. Under appropriate conditions $\{x^{(r)}\}$ will converge to a solution of the problem.

The penalized function $f(x) - R(x, \lambda)$ can be regarded as a special case of such a generalized Lagrangian, λ taking the role of the variable y. (The notation y will be henceforth reserved for the classic Lagrangian multiplier). However, no duality theory will be established or required: for the special cases considered any sequence $\{\lambda^{(r)}\}$ which increases to infinity will lead to a solution of the problem. Of course, one can still investigate whether any particular sequence is advantageous, in the sense of minimizing computation time. For example, it may not be advantageous to let $\{\lambda^{(r)}\}$ increase rapidly, or the solution $x^{(r)}$ for the rth stage may constitute a poor trial solution for the next stage. The consequent increase in the amount of computation needed per stage may then more than offset the decrease in the number of stages required for a given precision.

Heuristic rules sometimes employed with the penalty function (9.45) are, to choose the $\lambda_j^{(r+1)}$ so that $\lambda_j^{(r+1)}(g_j(x^{(r)}) - b_j)$ is roughly constant in j, and to increase the $\lambda_j^{(r)}$ at an exponential rate in r.

Fiacco and McCormick (1968), who were among the pioneers of the sequential approach, have a technique in which they separate the inequality and equality constraints, say

$$g_j(x) \leqslant 0 \quad (j = 1, 2, ..., m'),$$

$$g_j(x) = 0 \quad (j = m' + 1, ..., m),$$

and then work with the penalized function

$$\phi(x, \lambda) = f(x) + \lambda^{-2} \sum_1^{m'} g_j(x)^{-1} - \lambda \sum_{m'+1}^{m} g_j(x)^2.$$

Here λ is a scalar which tends to $+\infty$. It is plain that the part of the penalty function associated with the equality constraints shows the behaviour required by (9.43) in the limit, but the part associated with the inequality constraints in fact does not.

However, the notion is that if the first trial solution is in the region $\mathscr{X}_{\text{ineq}}$ specified by the inequalities, and so feasible as far as the inequality constraints are concerned, and the solution point is varied *continuously*, then it will never leave $\mathscr{X}_{\text{ineq}}$. For, no matter how large λ is, $\lambda^{-2} \sum g_j(x)^{-1}$ will tend to $-\infty$ as x tends to the boundary of $\mathscr{X}_{\text{ineq}}$ from within. There is thus a barrier at the boundary of $\mathscr{X}_{\text{ineq}}$ which can never be pierced by maximum-seeking methods which rely on a virtually continuous variation of the trial solution.

Despite the rather special character of its penalty function, the method seems to work well, judging from the considerable computing experience which has now been accumulated with it.

Exercise

(9.7) Let x^λ denote the value maximizing $f(x) - R(x, \lambda)$, and $\bar{x}$ denote the solution to the constrained problem. Suppose, moreover, that x^λ is feasible. Show then that

$$0 \leqslant f(\bar{x}) - f(x^\lambda) \leqslant \max_{x \in \mathscr{F}} R(x, \lambda) - R(x^\lambda, \lambda).$$

It will usually be the difference $f(\bar{x}) - f(x^\lambda)$ which is the important quantity. However, with appropriate conditions on f, these inequalities will also imply bounds on $\bar{x} - x^\lambda$.

9.5 CONVERGENCE RESULTS FOR SEQUENTIAL UNCONSTRAINED MAXIMIZATION PROCEDURES

It is helpful to consider a moderately specific problem, so we consider the case of inequality constraints: the problem $P(b)$ of maximizing $f(x)$ in $\mathscr{X}$

subject to $g(x) \leqslant b$. This includes the case of equality constraints, since a pair of scalar inequalities

$$\left. \begin{array}{r} g_1(x) \leqslant b_1, \\ -g_1(x) \leqslant -b_1 \end{array} \right\} \tag{9.47}$$

implies the equality $g_1(x) = b_1$. The notations $x(b)$, $U(b)$ and $\mathscr{X}(b)$ introduced in Section (2.2) will be used for solution, maximum value and feasible set, although the theorem below is stated for a fixed value of b.

Let x^λ be a value maximizing the penalized function

$$\phi(x, \lambda) = f(x) - Q(g(x) - b, \lambda) \tag{9.48}$$

in $\mathscr{X}$. It is assumed that Q and U have the following properties:

(i) $Q \geqslant 0$.

(ii) A positive constant K exists such that

$$Q(\delta, \lambda) \geqslant K \quad \text{for } \delta \nleqslant 0.$$

(iii)

$$\lim_{\lambda \to +\infty} Q(\delta, \lambda) = \left\{ \begin{array}{ll} +\infty & (\delta \nleqslant 0), \\ 0 & (\delta \leqslant 0). \end{array} \right. \tag{9.49}$$

(iv)

$$Q(\delta', \lambda) \geqslant Q(\delta, \lambda) \quad \text{for } \delta' \geqslant \delta. \tag{9.50}$$

(v) There is a function ψ such that, for the fixed value b and variable δ

$$|U(b) - U(b + \delta)| \leqslant \psi(\delta). \tag{9.51}$$

(vi)

$$\psi(\delta) \to 0 \quad \text{as } |\delta| \to 0. \tag{9.52}$$

(vii) $\psi(\delta)$ is of arbitrarily slower rate of growth in δ then $Q(\delta, \lambda)$ for δ positive and λ sufficiently large in the sense that

$$Q(\delta, \lambda) \leqslant Q(\varepsilon, \lambda) + \psi(\delta_+) \tag{9.53}$$

(for fixed $\varepsilon \leqslant 0$) implies that $|\delta_+| < d(\theta)$ if $\lambda_j \geqslant \theta$ for all components λ_j of λ, where $d(\theta) \to 0$ as $\theta \to +\infty$.

(viii) There exists an $\xi \geqslant 0$ such that $\mathscr{X}(b - \xi)$ is non-void.

Theorem (9.3)

If conditions (i)–(viii) are fulfilled, then a θ can be found such that $\lambda_j \geqslant \theta$ (all j) implies that x^λ is feasible, and that $f(x^\lambda)$ differs arbitrarily little from $U(b)$.

For definiteness, the reader can think of Q as the penalty function defined in (9.45).

Conditions (i)–(iv) are obvious requirements for a penalty function (with the possible exception of (ii): see Exercise (9.8)). No convexity or regularity conditions have been imposed on f, g or $\mathscr{X}$, save the conditions required of

$U(b)$ in (v)–(vii), which demand, roughly, that U shall not change too rapidly in the neighbourhood of the particular b value considered. This requirement seems best left as a condition on $U(b)$, since there are so many quite different sets of conditions on f, g and $\mathcal{X}$ that would imply it.

The final condition (viii) would seem to exclude equality constraints. However, if, for example, the inequality pair (9.47) were replaced by

$$g_1(x) \leqslant b_1 + \beta,$$

$$-g_1(x) \leqslant -b_1 + \beta,$$

with $\beta > 0$, then the form of the constraint no longer violates condition (viii), and the implied relation $-\beta \leqslant g_1(x) - b_1 \leqslant \beta$ can be made to approach an equality constraint as closely as desired, by choice of β small enough.

To prove the result, note first that

$$Q(g(x^\lambda) - b, \lambda) - Q(g(x) - b, \lambda) \leqslant f(x^\lambda) - f(x) \quad (x \in \mathcal{X}), \qquad (9.54)$$

since x^λ maximizes $\phi(x, \lambda)$ in $\mathcal{X}$. Note also that x^λ maximizes $f(x)$ subject to

$$g(x) \leqslant g(x^\lambda), \qquad (9.55)$$

so it can be asserted that

$$f(x^\lambda) = U(g(x^\lambda)). \qquad (9.56)$$

To see this, observe that, if x satisfies (9.55), then (9.55) together with (9.50) and (9.54) imply that $f(x) \leqslant f(x^\lambda)$.

Now set

$$\delta = g(x^\lambda) - b, \qquad (9.57)$$

so that $|\delta_+|$ measures the "degree of infeasibility" of x^λ. Setting $x = x(b)$ in (9.54), one deduces

$$
\begin{aligned}
Q(\delta, \lambda) - Q(g(x(b)) - b, \lambda) &\leqslant f(x^\lambda) - f(x(b)) \\
&= U(g(x^\lambda)) - U(b) \\
&= U(b + \delta) - U(b) \\
&\leqslant U(b + \delta_+) - U(b) \leqslant \psi(\delta_+).
\end{aligned} \qquad (9.58)
$$

Thus

$$Q(\delta, \lambda) \leqslant Q(g(x(b)) - b, \lambda) + \psi(\delta_+). \qquad (9.59)$$

Appealing to conditions (vi) and (vii), one sees that (9.59) implies that, for λ large enough (in the elementwise sense used in condition (vii)), $|\delta_+|$ and $\psi(\delta_+)$ can be made arbitrarily small, and x^λ brought arbitrarily near to feasibility.

Now set $x = x(b - \xi)$ (which exists by (viii)) in (9.54). Then

$$Q(\delta, \lambda) - Q(-\xi, \lambda) \leqslant Q(\delta, \lambda) - Q(g(x(b - \xi)) - b, \lambda)$$
$$\leqslant f(x^\lambda) - f(x(b - \xi))$$
$$= U(b + \delta) - U(b - \xi)$$
$$\leqslant U(b + \delta_+) - U(b - \xi)$$
$$\leqslant \psi(\delta_+) + \psi(-\xi), \tag{9.60}$$

so that

$$0 \leqslant Q(\delta, \lambda) \leqslant \psi(\delta_+) + \psi(-\xi) + Q(-\xi, \lambda). \tag{9.61}$$

Now, ξ can be chosen strictly positive, but small enough that $\psi(-\xi)$ is arbitrarily small. By the conclusion from (9.59), λ can then be chosen large enough that $\psi(\delta_+)$ is arbitrarily small. The same is true of $Q(-\xi, \lambda)$, by (9.49). Thus $Q(\delta, \lambda)$ can be made arbitrarily small by choice of sufficiently large λ. By conditions (ii) and (iii) this implies that $\delta \leqslant 0$, so that x^λ is in fact feasible.

Thus, for sufficiently large λ,

$$f(x^\lambda) \leqslant U(b) \tag{9.62}$$

while (9.60) implies that

$$f(x^\lambda) \geqslant f(x(b - \xi)) + Q(\delta, \lambda) - Q(-\xi, \lambda)$$
$$\geqslant U(b - \xi) - Q(-\xi, \lambda)$$
$$\geqslant U(b) - \psi(-\xi) - Q(-\xi, \lambda). \tag{9.63}$$

This last expression can be made to differ arbitrarily little from $U(b)$, so that the final assertion of the theorem follows from (9.62) and (9.63).

Exercises

(9.8) Note that a penalty function such as (9.46), $\sum \lambda_j(\delta_j)_+$, will not satisfy condition (ii). In such a case (9.61) will then imply "almost-feasibility" of x^λ rather than feasibility; i.e. not that δ_+ is zero, but that it is arbitrarily small, a fact already established from (9.59). However, (9.61) and (9.63) still hold, and, in place of (9.62), from (9.56)

$$f(x^\lambda) \leqslant U(b) + \psi(\delta_+)$$

so that $f(x^\lambda)$ still converges to $U(b)$. Find a weakened version of (ii) which supports this argument.

(9.9) If all functions involved are differentiable, and x^λ is interior to $\mathscr{X}$, then x^λ will be a stationary point of $\phi(x, \lambda)$, and

$$f_x - Q_\delta g_x = 0$$

at $x = x^\lambda$. One is thus led to suspect that the vector of differentials $\partial Q(\delta, \lambda)/\partial \delta$ (with $\delta = g(x^\lambda) - b$) might converge, with increasing λ, to the vector of classic Lagrangian multipliers $y = \partial U(b)/\partial b$, if this exists. This convergence, often rapid, is observed in numerical examples. Establish convergence under suitable conditions.

(9.10) Show that, if $x(b)$ is unique and f continuous, then $x^\lambda \to x(b)$.

CHAPTER 10

Vector Maximization Problems

An important and natural generalization of previous cases is that where $f(x)$, the function to be "maximized", is itself a vector.

In discussing the vector problem, one is led to mention the concepts of game theory, although we certainly do not attempt a serious treatment of these at any level—the subject is too individual and complex. The ideas which we particularly attempt to cover are those related to previous work: efficient points, the min–max theorem and its relation to Lagrangian theory, and, as a special case of this, the relation of the two-person zero-sum matrix game to linear programming.

10.1 THE VECTOR MAXIMUM PROBLEM: EFFICIENT POINTS

The problem of maximizing a function $f(x)$, subject to $x \in \mathscr{X}$, and perhaps to other constraints as well, has hitherto been the main concern. Consider, now, the case where f is itself a vector, so that it is desired to choose x in the feasible set in such a way as to make all components $f_1, f_2, ..., f_N$ of f large. For example, if x again describes an allocation of resources to various activities, then $f_1, f_2, ..., f_N$ may represent the resultant profit to each of N different agents involved—stock-holders in various companies, the state, the holders of concessions, etc. It is imagined that there is someone who has an interest in maximizing $f_i(x)$ for every $i = 1, 2, ..., N$.

Let $x \in \mathscr{X}$ be the only constraint. Now, it will rarely happen that all $f_i(x)$ will achieve their maximum value in $\mathscr{X}$ at a common point $\bar{x}$. In other words, there will rarely be some one allocation that suits everybody best. Usually the different $f_i(x)$ will attain their maxima at distinct x values, so that different people involved have different ideal allocations, and so have conflicting interests. This conflict of interest is a new and interesting feature: it has the implication that the vector maximum problem is not even properly posed, let alone solved.

It is not properly posed because there is no criterion indicating whether, for example, a change in x that increases $f_1(x)$ at the expense of a decrease in $f_2(x)$, by stated amounts, should be regarded as advantageous. Until such secondary criteria are supplied, there is no means of locating the balance-point in the conflict of interests.

What one can do, however, is to locate a class of points within which the solution to the vector maximum problem, however posed, must lie. Suppose that

$$f(x) > f(x'), \qquad (10.1)$$

this being understood to imply that

$$f_i(x) \geqslant f(x_i') \quad (i = 1, 2, ..., N), \qquad (10.2)$$

with strict inequality for at least one i. Then clearly x is preferable to x', since, in terms of the example, it is an allocation which leaves nobody worse off, and actually improves the return for somebody. x will be said to *dominate* x' if (10.1) holds. If $\bar{x}$ is not dominated by any x in $\mathscr{X}$, then $\bar{x}$ will be termed *efficient*. That is, $\bar{x}$ is efficient if there exists no x in $\mathscr{X}$ such that

$$f(x) > f(\bar{x}). \qquad (10.3)$$

Clearly, attention can be restricted to the set $\mathscr{E}$ of efficient points. In the absence of other criteria for narrowing down preferences further, it is the set that is analogous to the set of absolutely maximizing points in $\mathscr{X}$ for the case of scalar f. The efficient points correspond to the admissible strategies of decision theory; see Section (10.7).

There is a characterization of the efficient points which is very satisfying intuitively.

Theorem (10.1)

Suppose that $f(x)$ is bounded above $(x \in \mathscr{X})$, and that the set $f(\mathscr{X}) - R_+^N$ consisting of points of the form $f(x) - t$ $(x \in \mathscr{X}, t \geqslant 0)$ is convex. Then for any $\bar{x}$ in the efficient set $\mathscr{E}$ there is a vector $w > 0$ such that $\bar{x}$ maximizes $w^T f(x)$ in $\mathscr{E}$.

That is, the efficient points can be generated by maximizing a scalar utility function $w^T f(x)$, in which the individual utilities $f_i(x)$ are combined linearly, with weightings w_i.

The statement that $\bar{x}$ is efficient amounts to the assertion that $f(\bar{x})$ is a boundary point of the convex set $f(\mathscr{X}) - R_+^N$. Hence, there exists a supporting hyperplane at $f(\bar{x})$, i.e. a non-zero vector w such that

$$w^T(f(x) - t) \leqslant w^T f(\bar{x}) \qquad (10.4)$$

for $x \in \mathscr{X}$, $t \geqslant 0$ (see Fig. 10.1). Taking $x = \bar{x}$ one finds that $w^T t \geqslant 0$, or $w > 0$ (since $w \neq 0$); taking $t = 0$, one sees that $\bar{x}$ maximizes $w^T f(x)$ in $\mathscr{X}$. The theorem is thus proved.

The convexity condition on $f(\mathcal{X}) - R_+^N$ ensures that no efficient point is dominated by a mixture of other points.

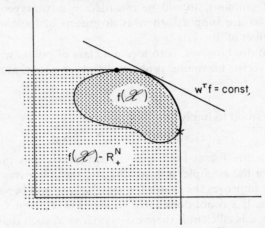

Figure 10.1 An illustration of the plane case $N = 2$. The set $f(\mathcal{X})$ is heavily shaded, the total shaded area and its continuation constitutes the set $f(\mathcal{X}) - R_+^N$. The heavily drawn section of the boundary constitutes the set of efficient points; a supporting hyperplane has been drawn at one of these

Since the vector maximization problem is now seen as essentially reducible to the scalar problem, the Lagrangian approach to the constrained vector case becomes clear: one operates with $w^{\mathrm{T}}f(x) - y^{\mathrm{T}}g(x)$ as we operated with $wf(x) - y^{\mathrm{T}}g(x)$ before. However, rather than seeing the calculation as a reduction to the scalar case followed by application of familiar Lagrangian methods, it is preferable to return to basic principles, and derive the two sets of multipliers simultaneously.

So, consider the location of efficient points for $f(x)$ in $\mathcal{X}$, subject to $b - g(x) \in \mathcal{C}$. Suppose that the set $(g(\mathcal{X}) + \mathcal{C}, f(\mathcal{X}) - R_+^N)$ is convex; that $\bar{x}$ is feasible and efficient, and $f(\bar{x}), g(\bar{x})$ finite. Then, by a familiar appeal to the supporting hyperplane theorem, it is deduced that there exist vectors w (in R_+^N) and y (in $\mathcal{C}^*$) such that $w^{\mathrm{T}}f(x) - y^{\mathrm{T}}g(x)$ is maximal in $\mathcal{X}$ at $\bar{x}$. The reader is left to deduce conditions under which $w > 0$, and to determine differential ("price") interpretations of w, y when these are possible. As before, if $\mathcal{C}$ is a convex cone, then the relation

$$y^{\mathrm{T}}(b - g(\bar{x})) = 0 \qquad (10.5)$$

holds.

Exercises

(10.1) The "conflict of interests" may be a conflict within an individual just as well as between individuals. For example, in the stochastic allocation problem of Section (6.2) one is trying to assure both large expectation of profit and small risk: these demands may well be in conflict. Reconcile the approach taken there with the considerations of this section.

(10.2) Show that $f(\mathscr{X}) - R_+^N$ is convex if $\mathscr{X}$ is convex and the functions $f_i(x)$ individually concave.

(10.3) Returning to the unconstrained case, one naturally asks whether the converse to Theorem (10.1) holds: that the x maximizing $w^T f(x)$ in $\mathscr{X}$ is efficient for any $w > 0$. There must be reservations about cases where w has some zero components; i.e. where the supporting hyperplane may be parallel to a co-ordinate axis. In Figure (10.2) are illustrated some two-dimensional

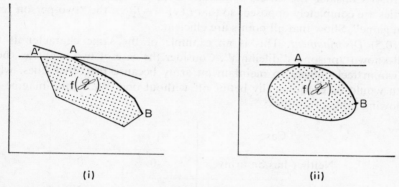

Figure 10.2   In the case (i) of polygonal $f(\mathscr{X})$ all efficient points (on the boundary sequent AB) have supporting hyperplanes with finite and strictly negative slope. In the case (ii) this is not true of the outermost efficient points, A and B

cases ($N = 2$): the efficient points correspond to the north-east section of the boundary of $f(x)$, marked AB in the figures. Supporting hyperplanes to $f(\mathscr{X}) - R_+^N$ are also supporting hyperplanes to $f(\mathscr{X})$ at these boundary points. For the polygonal case (i) all points in AB can obviously be attained as contact points of hyperplanes of strictly negative but finite slope (i.e. these correspond to x-values maximizing $w^T f(x)$ for some $w \gg 0$). If horizontal supporting hyperplanes (with $w_1 = 0$) are considered then all points on the side $A'A$ also become eligible as contact points (i.e. maximize $w^T f(x)$) despite the fact that of these only A corresponds to an efficient point. In the case (ii), it is nevertheless necessary to consider a hyperplane with $w_1 = 0$ if the point A is to be included at all, i.e. if the corresponding x-value is to maximize $w^T f(x)$ for some $w > 0$.

On the basis of these observations, prove the following assertions. Convexity and boundedness of $f(\mathcal{X})$ may be assumed throughout.

(i) Suppose that $w \gg 0$ and that $\bar{x}$ maximizes $w^T f(x)$ in $\mathcal{X}$. Then $\bar{x}$ is efficient.

(ii) Let $\mathcal{F}$ be the closure of the set of points obtained in (i): x-values maximizing $w^T f(x)$ for some $w \gg 0$, and their limits. Then $\mathcal{F}$ is exactly the class of efficient points $\mathcal{E}$.

(iii) Let $\mathcal{G}$ be the set of x-values maximizing $w^T f(x)$ for some $w > 0$. Then this is the set of $\bar{x}$ for which there exists no x in $\mathcal{X}$ such that $f(x) \gg f(\bar{x})$. This characterizing property is weaker than the property of efficiency, as we see from (10.3).

(10.4) Consider the case $N = 2$, and suppose that the interests of the parties are completely opposed, so that $f_2(x) = -f_1(x)$; the "two-person zero-sum game". Show that all points are efficient.

(10.5) *Disarmament.* This is an example of the same character as the well-known "prisoners' dilemma". Consider the case of two countries, both of whom feel obliged to maintain an army because the other does, while both would be economically better off without one. We could imagine the following table of utilities:

Case	f_1	f_2
Neither has an army	0	0
1 has, 2 does not	1	-2
2 has, 1 does not	-2	1
Both have armies	-1	-1

Verify that all situations are efficient, save the last.

This example will be considered again in Section (10.3).

10.2 EXCHANGE EQUILIBRIA

Suppose that the N "players" are N economic units; for example, different countries, or different sectors of a single isolated economy. Suppose that there is a trade in several commodities between the units, and that unit i exports a net amount x_{ij} of commodity j. A negative value of x_{ij} thus corresponds to a net import. There is no need to specify the destination of the export (or origin of the import); it is required only that there be no net import into the system as a whole, i.e.

$$\sum_{i=1}^{N} x_{ij} \geqslant 0. \tag{10.6}$$

Denote the column vector $(x_{i1}, x_{i2}, \ldots)$ by x_i: the "export vector" for the ith unit. Relations (10.6) can then be summarized as

$$\sum_{1}^{N} x_i \geqslant 0. \tag{10.7}$$

The utility of this whole commerce to unit i is presumably a function only of the resources it is left with: $b_i - x_i$, if b_i is its vector of naturally occurring resources. For fixed b this can then be written as a function $f_i(x_i)$; one has then the vector maximization problem of choosing the pattern of commerce $x = (x_1, x_2, \ldots, x_N)$ in such a way as to maximize the utility vector $(f_1(x_1), f_2(x_2), \ldots, f_N(x_N))$ subject to (10.7).

Presumably only efficient points need be considered as possible solutions of this problem, for if a change in x would benefit unit 1, say, and leave nobody else worse off, then unit 1 could induce others to change by offering them a share of its increased utility. Actually, this is a dangerous line of argument to open, because it implies a modification of the basic utility functions and also implies the existence of some mechanism for transferring utility between units. The notions of coalition and cooperation lead to subtle and difficult analysis. A few ideas are mentioned in passing in Section (10.4).

If the f_i are individually concave for all values of their real vector argument, then all convexity assumptions are fulfilled, and one can assert the existence of non-negative vectors w and y such that a given efficient point $\bar{x}$ maximizes

$$L(x, y) = \sum w_i f_i(x_i) + y^T \sum x_i \tag{10.8}$$

absolutely, over all real x in a vector space of appropriate dimension. If the maximum is attained at a stationary value of L then

$$-\frac{\partial f_i}{\partial x_{ij}} = \frac{y_j}{w_i}. \tag{10.9}$$

This would be the marginal price that unit i would be willing to pay for an additional amount of commodity j. It is thus seen that y_j is effectively a "world price" for commodity j (in the marginal sense) and w_i an exchange rate for unit i. Relation (10.5) becomes in this case

$$y^T \sum \bar{x}_i = 0; \tag{10.10}$$

a statement that the surplus of the system has no value on prices y.

8

The exchange rates w_i can only be determined by requiring other properties of the pattern of trade. A criterion sometimes applied is that the stronger version of (10.10),

$$y^T \bar{x}_i = 0 \quad (i = 1, 2, ..., N) \tag{10.11}$$

should hold, i.e. that each unit should individually be in trade-balance on the basis of prices y. However, since the y_j are only marginal prices, the justification for such a requirement is not clear.

Another requirement might be that economic units should cooperate rather than compete. Suppose utilities are additive in the sense that, if a pattern x is adopted, then the total utility for units in a set A is $\sum_A f_i(x_i)$ and this total utility can be shared among members of the set in arbitrary proportions. Then maximum cooperation would be in everybody's interest, i.e. x should be chosen to maximize $\sum_1^N f_i(x_i)$. This is true in the sense that a distribution of this total utility can be found which will leave every unit at least no worse off than if some other given pattern x' had been adopted. Maximization of $\sum f_i$ corresponds, in view of the conclusion from (10.9), to the adoption of a common exchange rate for all units.

Exercises

(10.6) Relation (10.9) states that the matrix $(\theta_{ij}) = (\partial f_i / \partial x_{ij})$ is of rank one. Prove the necessity for this condition (assuming differentiability, etc.) by requiring optimality of trade between any two given units in any two given commodities, without appeal to Lagrangian methods.

(10.7) Find an analogue of relation (10.9) for the case where L is not maximized at a stationary point.

10.3 EQUILIBRIUM POINTS OF AN N-PERSON GAME

A vector problem of rather special structure is the *N-person game*. This is a situation similar to that of Section (10.1), in that $f_i(x)$ is interpreted as a return to player i if x is the course of play in the game. However, there are additional assumptions concerning the degree of control each player can exert on the value of x. This question of control was not made explicit in Section (10.1); it was merely assumed that a mechanism existed which followed some principle of maximal mutual advantage to seek out an appropriate x, presumably one of the efficient points.

In the game formulation the assumption is added that x can be partitioned into components (with vector or more general values) $x_1, x_2, ..., x_N$, and that player i has complete control over the value of x_i. The component x_i thus describes player i's rules for his moves; that part of the play over which he has choice. It will be termed his *strategy*. He will attempt to choose x_i so as to maximize f_i. The aims of different players will usually be in conflict, and the fact that each player has an assigned component of x under his control makes this element of conflict or competition even more explicit than it was in Section (10.1).

A "game" can mean a relatively serious matter: e.g. bargaining between employers and workers, or the disposition of attacking and defending forces in warfare.

Most games take place in time, evolving step by step. This dynamic aspect is one of the most interesting features, and probably must be considered if one is to achieve a satisfactory characterization of the "solution"of a game. However, there will not be space to discuss it here, and the conventional approach will be adopted, in which one treats a game as a relatively static matter, in which each player chooses his strategy x_i, and the game then proceeds and throws up the returns $f_i(x)$.

$f_i(x)$ can, for explicitness, be written $f_i(x_1, x_2, ..., x_n)$ but a more convenient notation is $f_i(x_i, r_i)$, where r_i denotes the strategies $x_1, x_2, ..., x_{i-1}, x_{i+1}, ..., x_N$ of all players but i, collectively.

The properties that would characterize the solution of a game are not immediately clear. However, one appealing notion is that of an *equilibrium point*. The point $\bar{x}$ is said to be an equilibrium point if

$$f_i(\bar{x}_i, \bar{r}_i) \geqslant f(x_i, \bar{r}_i) \quad (i = 1, 2, ..., N). \tag{10.12}$$

Here x_i is any other conceivable strategy for player i, and $\bar{r}_i$ is the value of r_i when $x = \bar{x}$. Relation (10.12) states that no player has any incentive to change his strategy if all the others hold theirs fixed.

What this concept excludes is the possibility of cooperation: that there might be an incentive to change if several players were to decide jointly to change their existing strategies to new values. If several players are acting jointly like this, they form a coalition. In considering a change of strategy, the coalition would also have to foresee subsequent reactions on the part of other players, which is virtually a return to considering the game as a dynamic problem. However, only the so-called *non-cooperative* games, will be considered, in which each participant plays on his own, and for which an equilibrium point $\bar{x}$ as characterized above has an obvious stability.

One of the central topics of game theory is the determination of conditions under which equilibrium points exist. The existence of an equilibrium point implies the celebrated min–max theorem as a special case, and this, as will be seen, ties up with the earlier work on Lagrangian theory.

Theorem (10.2)

Suppose that x_i can be chosen in a set $\mathscr{X}_i$ which is bounded, closed, non-empty and convex, and that the value of x_i maximizing $f_i(x_i, r_i)$ in $\mathscr{X}_i$ is continuous in r_i. Then at least one equilibrium point exists.

For, consider a strategy $\bar{x}$, and a strategy

$$\tilde{x} = T(\bar{x})$$

derived from this by the requirement that $\tilde{x}_i$ should be the x_i-value maximizing $f_i(x_i, \bar{r}_i)$. Then, by the assumptions of the theorem, $x \to T(x)$ is a continuous mapping of the bounded, closed, non-empty point-set

$$(\mathscr{X}_1, \mathscr{X}_2, ..., \mathscr{X}_N)$$

into itself. One can thus appeal to the Brouwer fixed-point theorem (see Lefschetz, 1949, or the Appendix to Burger, 1959) and assert that the mapping has at least one fixed point $\bar{x}$, such that

$$\bar{x} = T(\bar{x}).$$

By the construction of the mapping T, the point $\bar{x}$ is then an equilibrium point, and the theorem is proved.

It is a pity that one has to appeal to a non-elementary result, the Brouwer theorem, but a short, elementary proof of Theorem (10.2) does not seem to exist. (See, however, Exercise (10.9) for an elementary proof in a special case.) One attractive type of proof relies on an iterative argument, in which a sequence of "improved" strategies $\{x^{(n)}\}$, generated by $x^{(n+1)} = T(x^{(n)})$, or some modification of this, is studied directly. Such an approach is appealing, because it offers a method of actually finding equilibrium points, by the successive improvement of an initial strategy, and a method which may moreover approximate the mechanism by which good strategies are evolved in real life. However, proofs of convergence of the sequence $\{x^{(n)}\}$ are difficult and delicate (see Karlin, 1959, p. 179).

A condition which ensures continuity of T is the following:

Corollary

Suppose that, in addition to the assumptions on $\mathscr{X}_i$ $(i = 1, 2, ..., N)$ of Theorem (10.2), f_i is concave in x_i for given r_i, and continuous in all arguments. Then the theorem holds.

For, if f_i is strictly concave in x_i $(i = 1, 2, ..., N)$ the mapping T is readily verified to be continuous, so that the assertions of the theorem are true. If f_i is merely concave, it may be made strictly concave by the addition of a strictly concave function differing by an arbitrarily small amount from zero. The assertion then follows by a limiting argument.

A point can be efficient without being an equilibrium point, and vice versa. To prove the first assertion, consider the example of Exercise (10.4), for which all points were efficient, and yet which possesses an equilibrium only under special conditions. To prove the second assertion, consider the disarmament example of Exercise (10.5), for which the only equilibrium point is the only inefficient one: the case where both countries arm. Either side loses utility if it disarms unilaterally, so a situation in which both countries are armed is an equilibrium point. However, this point is not efficient, for both countries would be better off if they could agree to disarm simultaneously. On the other hand, universal disarmament is not an equilibrium point: either side could gain a temporary advantage by arming, and disarmament can be maintained only by a mutual decision.

These models of "social conflict" are bound, of course, to be over-simplified, and some of the dilemmas may be artificial products of an incomplete model. For example, the situation should often be formulated dynamically rather than statically—as a sequence of actions and reactions rather than as a choice of frozen stances. Then, there may be possibilities outside the model which would resolve the dilemma. For example, if one country were to conquer the other, this would constitute an "enforced coalition" within which disarmament could presumably be achieved, in the end.

Exercises

(10.8) Consider the two-person game of hide-and-seek. Suppose the hider may hide in one of s places, and the seeker is allowed to look only in one of these. The winner scores 1, the loser 0.

Show that an equilibrium strategy for each is the mixed one, in which the hider chooses a hiding place at random from the s possibilities with equal probabilities, and the seeker does likewise.

(10.9) Consider the following "learning sequence" convergent to the solution of Exercise (10.8). Suppose the game is repeated indefinitely, the hider (seeker) going to one of the places the seeker (hider) has visited least (most) frequently in the past. Show that the proportion of the time either player visits the various hiding places tends to the uniform distribution of Exercise (10.8) as the number of games played increases indefinitely.

(10.10) In Exercise (10.9) it was assumed that, after every game, each player learned which hiding-place the other had visited, whatever the outcome of the game. Suppose now that this is true only for those games in which "contact" has been made, and so which the seeker has won. The hider (seeker) must then use one of the hiding-places in which there has least (most) frequently been a contact. Convergence is now a stochastic matter, and much more difficult to prove.

(10.11) *Equilibrium of an oligopoly.* Suppose that N potential producers can supply a market with a particular commodity, and that, if the ith supplies an amount x_i, then his return is

$$f_i(x) = x_i p\left(\sum_1^N x_j\right) - c(x_i).$$

That is, his cost of production is $c(x_i)$, and the unit price received for the commodity is $p(\sum_1^N x_i)$, a function of the total supply. Each producer wishes to increase his share of the market, without producing so much that the general price falls to an uneconomic level.

Suppose $c(x_i)$ increasing, differentiable and convex in $x_i \geqslant 0$, and $p(\xi)$ decreasing, differentiable and concave in the interval $0 \leqslant \xi \leqslant \xi_0$, and equal to zero from ξ_0 onwards.

Show that there is exactly one equilibrium point: $x_i = \alpha$ $(i = 1, 2, ..., N)$, where $\alpha = 0$ if $c'(0) \geqslant p(0)$, or α equals the only root of

$$p(N\alpha) + \alpha p'(N\alpha) = c'(\alpha)$$

in $0 < \alpha < \xi_0/N$ otherwise.

(Hint: if α_j, α_k are two distinct non-zero values adopted by one of the x_i at equilibrium, show that

$$(\alpha_j - \alpha_k) p'(\sum x_i) = c'(\alpha_j) - c'(\alpha_k).)$$

The convexity assumptions on p and c are as questionable as they were in Section (8.1). Show that if these are replaced by the assumption of strictly monotone c', still only two non-zero values α_j are possible for the x_i at equilibrium. Other modifications in the direction of realism are possible, such as establishing a threshold on entry into production by setting $f_i(0) = M > 0$ (where M is the income a potential producer might make from engaging in some other activity), or of allowing cooperative behaviour of various shades of ethicality.

10.4 TWO-PERSON GAMES; THE MIN-MAX THEOREM; COMPLEMENTARY VARIATIONAL PRINCIPLES

Consider a game with two players, in which one player's gain is the other's loss, so that

$$f_2(x_1, x_2) = -f_1(x_1, x_2). \tag{10.13}$$

This is called a *two-person zero-sum game*, for obvious reasons. The notation (x, y) will be used instead of (x_1, x_2) and f_1 denoted simply by ϕ. Then an equilibrium point will now be a point $\bar{x}, \bar{y}$ for which

$$\phi(x, \bar{y}) \leqslant \phi(\bar{x}, \bar{y}) \leqslant \phi(\bar{x}, y) \tag{10.14}$$

for all x, y in the relevant sets $\mathscr{X}, \mathscr{Y}$. Such a point is termed a *saddle-point* of the function ϕ, also a saddle-point of the game, by extension of the term.

Theorem (10.3)

(*Equivalence*) *All saddle-points of a zero-sum two-person game are equivalent, in the sense that ϕ takes the same value at all of them.*

This common value, v, say, is termed the *value* of the game.

To prove the theorem, suppose that x', y' and x'', y'' are both saddle-points. Then

$$\phi(x', y') \leqslant \phi(x', y'') \leqslant \phi(x'', y''). \tag{10.15}$$

The reverse inequality follows in the same way, so that $\phi(x', y')$ and $\phi(x'', y'')$ are indeed equal.

In fact, note further that

$$\phi(x, y') \leqslant \phi(x', y') = \phi(x'', y'') \leqslant \phi(x'', y'). \tag{10.16}$$

The same proof yields the inequality

$$\phi(x'', y') \leqslant \phi(x'', y), \tag{10.17}$$

so that (x'', y') is also a saddle-point of ϕ.

Theorem (10.4)

(*Interchangeability*) *If (x', y'), (x'', y'') are saddle-points of a zero-sum, two-person game, then so are (x'', y') and (x', y'').*

The following consequence of the existence of a saddle-point gives an illuminating alternative partial characterization of an equilibrium solution.

Theorem (10.5)

If the game possesses a saddle-point, then

$$\max_{x} \min_{y} \phi(x, y) = \min_{y} \max_{x} \phi(x, y), \tag{10.18}$$

and the common value is the value of the game.

Here the maximum and minimum operations are applied over $\mathscr{X}$ and $\mathscr{Y}$ respectively. The interpretation is, that the right-hand member of (10.18) is the pay-off to player 1 if he declares his strategy (x) after player 2 has declared his (y); the other member is the pay-off for the reverse order of play. The theorem thus states that there is no disadvantage in declaring first (as there is usually, see (10.20) and Exercises (10.12)–(10.14).

To prove the result, note that for any x, y

$$\min_{y} \phi(x, y) \leqslant \phi(x, y) \leqslant \max_{x} \phi(x, y), \tag{10.19}$$

whence

$$\max_x \min_y \phi(x, y) \leqslant \min_y \max_x \phi(x, y). \tag{10.20}$$

On the other hand, if ϕ has a saddle-point $(\bar{x}, \bar{y})$, then

$$\min_y \max_x \phi(x, y) \leqslant \max_x \phi(x, \bar{y}) = \phi(\bar{x}, \bar{y}), \tag{10.21}$$

and, by the same argument,

$$\phi(\bar{x}, \bar{y}) \leqslant \max_x \min_y \phi(x, y). \tag{10.22}$$

Inequalities (10.20)–(10.22) imply that the two members of (10.18) both equal $\phi(\bar{x}, \bar{y})$. The converse to Theorem (10.5) is not true without further conditions.

Sufficient conditions for existence of a saddle-point follow from specialization of Theorem (10.2) and its corollary.

Theorem (10.6)

(*The min-max theorem*) *The two-person zero-sum game with pay-off $\phi(x, y)$ on respective strategy sets $\mathcal{X}, \mathcal{Y}$ has an equilibrium point (saddle-point) if (i) $\mathcal{X}, \mathcal{Y}$ are bounded, closed, non-empty and convex, (ii) $\phi(x, y)$ is concave in x for fixed y, convex in y for fixed x, and (iii) $\phi(x, y)$ is continuous in (x, y).*

The min-max theorem has applications to situations other than two-person games. It has already been seen in Exercise (3.1) that validity of the strong Lagrangian principle is equivalent to the commutativity of max and min operators in a certain case. To take a case which is at least seemingly more general, consider the Lagrangian form

$$L(x, y) = f(x) + y^{\mathrm{T}}(b - g(x))$$

for the problem $P_{\mathscr{C}}(b)$ in the case of $\mathscr{C}$ a cone. Let the domains of variation of x and y be understood to be $\mathscr{X}$ and $\mathscr{C}^*$ respectively. Then

$$\min_y L(x, y) = \begin{cases} f(x) & (x \in \mathscr{X}(b)), \\ -\infty & \text{otherwise,} \end{cases}$$

so that

$$\max_x \min_y L(x, y) = U(b)$$

at least if f is finite. But it is known from Theorem (3.3) that

$$U(b) = \min_y \max_x L(x, y)$$

if a one-point solution to the problem can be found by the strong Lagrangian principle. This is then a case where max and min commute, and where the solution $\bar{x}$ to the problem and the associated multiplier value y constitute a saddle-point of the Lagrangian form $L(x, y)$. It is true, then, that the Lagrangian theory of Chapter 3 could have been established by appeal to a max-min theorem. However, this more general approach would have been less adapted to the particular problem, of constrained maximization, and would have required appeal to an extraneous result: the fixed-point theorem.

It is interesting, nevertheless, that the constrained minimization problem can be viewed as a game, in which a free choice of x in $\mathscr{X}$ is countered by the existence of an "opponent" who chooses y (the shadow price vector, in the allocation problem) in a fashion most disadvantageous to the optimizer. The interpretation is sometimes reversible, as will be seen in Section (10.6).

In the next section another application is given of the min-max theorem. Yet a further application is provided by the pairs of complementary variational principles of physics, discussed in Section (6.5). Suppose that x and y denote two sets of physical variables, whose values are determined by the criterion that some function $\phi(x, y)$ shall be simultaneously maximal with respect to x and minimal with respect to y. In order that these extremal operations shall have a meaning when applied simultaneously, it will be assumed that they commute, as in (10.18). This they will do if ϕ obeys the conditions of the min-max theorem, for example. Let $x(y)$ denote a value of x that maximizes $\phi(x, y)$ for a given y, and $y(x)$ a minimizing value of y for given x, so that

$$\phi(x, y(x)) \leqslant \bar{\phi} \leqslant \phi(x(y), y), \tag{10.23}$$

where $\bar{\phi}$ is the max-min = min-max value defined in (10.18). Furthermore, because a saddle-point exists and equality holds in (10.18), there is an x and a y for which equality holds throughout in (10.23), the x-value necessarily maximizing $\phi(x, y(x))$ and the y value minimizing $\phi(x(y), y)$. These are the complementary variational principles: that x shall maximize $J(x) = \phi(x, y(x))$ and y shall minimize $K(y) = \phi(x(y), y)$, the extreme values of J and K being equal. These "principles" follow fairly trivially from the existence of a compound principle, that ϕ possesses a saddle-point at which it is simultaneously maximal and minimal. However, much work and insight is required before a particular body of physical theory acquires a form which reveals the validity of such general principles. See Exercises (10.16)–(10.18).

Exercises
(10.12) Consider a version of the game of Exercise (10.8) in which mixed strategies are *not* permitted, so that each player's rule of play amounts to definite choice of a place in which to hide or seek, rather than of a distribution over places. That is, x and y take values in a common set of r points, and effectively,

$$\phi(x, y) = \begin{cases} 1 & (x = y), \\ -1 & (x \neq y). \end{cases}$$

Show that the min-max theorem does not hold for this ϕ: interpret this in terms of the importance of "order-of-play" for the unrandomized game.

(10.13) Player 1 backs one of two horses in a race up to a unit amount: player 2 attempts to thwart him by doctoring one of the two horses so that it cannot win. If x is the difference in stake placed on horses 1 and 2, then, effectively

$$\left.\begin{array}{rl} \phi(x, 1) = & x \\ \phi(x, 2) = & -x \end{array}\right\} \quad (-1 \leqslant x \leqslant 1),$$

where the two values of y correspond to the case where player 2 nobbles horse 2 or horse 1. Show that the min-max theorem is not satisfied, and relate this to the importance of "order-of-play". (A rather more general version of this situation: players 1 and 2 drive a car jointly, one operating the steering-wheel and the other the accelerator, and they wish to reach different places.)

(10.14) Following on from Exercise (10.13), note that the min-max theorem will generally not be satisfied for

$$\phi(x, y) = a_y^{\mathrm{T}} x + b_y,$$

where y takes a finite set of values, and x varies in some set in R^n.

(10.15) Suppose that ϕ satisfies the convexity-concavity assumptions of Theorem (10.6). By appealing to Theorem (10.4), show that convex sets $\mathcal{X}_0, \mathcal{Y}_0$ exist such that (x, y) is a saddle-point of ϕ if $x \in \mathcal{X}_0, y \in \mathcal{Y}_0$

(10.16) Suppose x and y are vectors of the same dimensionality, with elements x_r, y_r $(r = 1, 2, \ldots, n)$ which simultaneously maximize and minimize an expression

$$I(x, y) = \sum_r W(x_r, y_r, r) - y^{\mathrm{T}} A x.$$

Find conditions on W for $\max_x$ and $\min_y$ to commute. Note that if extrema with respect to either variable for fixed values of the other are attained at stationary points of I then x, y obey

$$\sum_j y_i a_{jr} = \frac{\partial W(x_r, y_r, r)}{\partial x_r}, \tag{10.24}$$

$$\sum_j a_{rj} x_j = \frac{\partial W(x_r, y_r, r)}{\partial y_r} \tag{10.25}$$

Note that the function $J(x)$ defined in the text is $I(x, y(x))$ with $y(x)$ determined by (10.25).

(10.17) Note the continuous space analogue of Exercise (10.16):

$$I = \int W(x_r, y_r, r) \, dr - \int y_r A x_r \, dr,$$

where A is a linear operator with adjoint A^T. Under appropriate conditions the "fields" x_r, y_r then satisfy

$$Ax = \frac{\partial W}{\partial y},$$ (10.26)

$$A^T y = \frac{\partial W}{\partial x},$$ (10.27)

where x, y, W are now used simply to denote the point values $x_r, y_r, W(x_r, y_r, r)$ with the co-ordinate argument r understood. The complementary variational principles are that x shall maximize $I(x, y)$ with y determined in terms of x by (10.26), and y shall minimize $I(x, y)$ with x determined in terms of y by (10.27). (See Noble (1964) who develops these results for the case when W has the appropriate convexity/concavity properties locally, and minima and maxima are also considered as local.)

(10.18) (Arthurs and Robinson, (1968)) The energy of an electrostatic field in a space with co-ordinate r can be expressed

$$I = \max_{\phi} \min_{U} \frac{1}{4\pi} \int \left[\frac{1}{2\kappa} |U|^2 + 4\pi\rho\phi - U.\operatorname{grad}\phi \right] dr,$$

where ϕ is the electrostatic potential, U the electrostatic force field, κ (>0) the dielectric parameter and ρ the charge density. κ and ρ are prescribed and will usually be r-dependent. The operator grad is adjoint to the operator $-\operatorname{div}$ in that

$$\int U.\operatorname{grad}\phi = -\int \phi\operatorname{div}U + \text{boundary terms},$$

where the "boundary terms" correspond to an integral over the surface of the space considered, which may be supposed zero under a variety of circumstances. (Note: if r and U have Cartesian components r_k and U_k, then gradϕ is a vector with components $\partial\phi/\partial r_k$, and $\operatorname{div}U = \sum_k \partial U_k/\partial r_k$.)

Is there a saddle-point in this case? Show that ϕ, U satisfy the following complementary variational principles, the quantity rendered extreme being the field energy in each case: ϕ maximizes

$$J(\phi) = \int \left[\rho\phi - \frac{\kappa}{8\pi} |\operatorname{grad}\phi|^2 \right] dr$$

(the Dirichlet principle), and U minimizes

$$K(U) = \frac{1}{8\pi} \int |U|^2/\kappa \, dr$$

subject to

$$\text{div } U + 4\pi\rho = 0$$

(the Thomson principle).

10.5 AN APPLICATION OF THE MIN-MAX THEOREM TO EXPERIMENTAL DESIGN

Suppose that some quantity v (say, the flexure of an aircraft wing) is nonlinearly dependent upon another, u (say, the wing loading) in such a way that one can represent the relation between observations as these two quantities by a polynomial regression

$$v = \sum_{j=0}^{r} \beta_j u^j + \varepsilon, \qquad (10.28)$$

where ε is a random deviation and β_j a regression coefficient.

Several pairs of observations $u(i), v(i)$ will normally be taken and the coefficients β_j estimated by least squares. Suppose now that the coefficient of the highest power of u, β_r, is of particular interest, because the term u^r represents the greatest source of nonlinearity in the relation. In such a case, one might demand that the estimate b_r of β_r should have the smallest possible sampling variance, because one will then obtain the most sensitive test of whether the term in u^r should be included in the regression or not.

Frequently there is an element of choice in the design of the experiment: the values $u(i)$ can be chosen by the experimenter. For instance, in the wing flexure example mentioned, the loadings applied could be chosen by the experimenter, if the test were a static one rather than in-flight. The question then arises, how should these values be chosen so as to minimize the sampling variance of b_r?

Theorem (10.7)

Suppose that errors $\varepsilon(i)$ are uncorrelated, with zero mean and constant variance, and that the u values may be chosen in the interval $(-1, 1)$. In order to minimize the sampling variance of b_r, the variable u should be assigned the values $u_1, u_2, \ldots, u_{r+1}$ in the proportions $1/2r, 1/r, 1/r, \ldots, 1/r, 1/2r$. Here $u_1, u_2, \ldots, u_{r+1}$ are the values, in increasing order, at which the polynomial

$$T_r(u) = \sum_{j=0}^{r} a_j u^j \qquad (10.29)$$

takes its extreme absolute value in the interval $(-1, 1)$, where T_r is the polynomial for which this extreme value is least, subject to $a_r = 1$.

As is well known in analysis, the polynomial thus characterized is the *Tchebichev polynomial*

$$T_r(u) = 2^{-r}\{[u + i(1 - u^2)^{\frac{1}{2}}]^r + [u - i(1 - u^2)^{\frac{1}{2}}]^r\}$$

$$= 2^{1-r}\cos[r\cos^{-1}(u)], \tag{10.30}$$

and the proof of Theorem (10.7) sheds a good deal of incidental light upon its nature.

The proof rests on the fact that, under the given assumptions on ε, the sampling variance of b_r is inversely proportional to

$$\min_a \sum_i \left(\sum_{j=0}^r a_j u(i)^j\right)^2, \tag{10.31}$$

the minimization being subject to $a_r = 1$. The aim is then to choose the $u(i)$ so as to maximize this quantity. However, the treatment is simplified if one considers rather the quantity

$$\min_a \phi(F, a) = \min_a \int_{-1}^1 (\sum a_j u^j)^2 \, dF(u), \tag{10.32}$$

where F is a distribution function, and maximizes expression (10.32) with respect to F. That is, the fact is being neglected that, if n observations are taken, then the distribution of u values in the experiment must be concentrated in quanta of n^{-1}.

The point of the assumption that u can take values only in $(-1, 1)$ is to ensure that the problem is non-degenerate. In practice, u values will be bounded for reasons of physical feasibility in both directions, and a standardizing linear transformation will bring these bounds to the values ± 1.

The fact that expression (10.32) is minimal with respect to $a_0, a_1, \ldots, a_{r-1}$ implies that

$$\int T_r(u) R(u) \, dF(u) = 0, \tag{10.33}$$

where T_r is given by (10.29) (although without, as yet, the characterization that follows (10.29)) and $R(u)$ is any polynomial of degree less than r.

Now, $\phi(F, a)$ has the properties required by the min-max theorem, allowing us to reverse the order of a-minimization and F-maximization. It thus follows that $a_0, a_1, \ldots, a_{r-1}$ must minimize

$$\max_F \int_{-1}^1 |T_r(u)|^2 \, dF(u) = \left[\max_{-1 \leqslant u \leqslant 1} |T_r(u)|\right]^2 = \Delta_r^2, \tag{10.34}$$

say. That is, T_r has indeed the min-max modulus characterization following (10.29) in the theorem, and the optimal u-distribution is concentrated on these values at which $T_r(u)$ attains its maximum modulus in $(-1, 1)$. If these values are denoted by u_k (in order) and the corresponding increments in F by p_k, then (10.33) becomes

$$\sum_k [\operatorname{sgn} T_r(u_k)] p_k R(u_k) = 0. \tag{10.35}$$

Now there cannot be more than $r+1$ values u_k: the $r-1$ turning points of T_r, and the endpoints of the interval, ± 1. We shall show that there are exactly $r+1$ such values, and that T_r alternates in sign as they are taken in order, so that (10.35) becomes

$$\sum_{k=1}^{r+1} (-)^k p_k R(u_k) = 0. \tag{10.36}$$

For, suppose that the sequence $p_k \operatorname{sgn} T_r(u_k)$ has fewer than r sign changes. Then a polynomial $R(u)$ can be found of degree less than r which changes sign with it, and so for which the sum in (10.35) is strictly positive. This would contradict (10.35), so the sequence does have at least r sign changes. This implies, since the number of terms is at most $r+1$, that there are exactly $r+1$ terms, with alternating sign.

All that remains is to prove the assignment of weights; that

$$p_k = \begin{cases} \dfrac{1}{2r} & (k = 1, r+1), \\[2mm] \dfrac{1}{r} & (1 < k < r+1). \end{cases} \tag{10.37}$$

Let

$$Q(u) = \prod_{j=1}^{r+1} (u-u_j),$$
$$Q_k(u) = \prod_{j \neq k} (u-u_j). \tag{10.38}$$

Taking $R(u) = Q_k(u) - u^r$, one sees from (10.36) that

$$p_k = \frac{(-)^k \operatorname{const}}{Q_k(u_k)} = \frac{(-)^k \operatorname{const}}{Q'(u_k)}. \tag{10.39}$$

Now, since $u_2, u_3, \ldots, u_r$ are double roots of

$$T_r(u)^2 = \Delta_r^2 \tag{10.40}$$

and single roots of

$$T_r'(u) = 0, \tag{10.41}$$

and u_1, u_{r+1} are simple roots of (10.40), one can take

$$Q(u) = r \frac{T_r(u)^2 - \Delta_r^2}{T_r'(u)}$$

so that

$$Q'(u) = r \left[2T_r - \frac{T_r''(T_r^2 - \Delta_r^2)}{(T_r')^2} \right]. \tag{10.42}$$

It follows from (10.40)–(10.42) that

$$Q'(u_k) = \text{const} \begin{cases} 2(-)^k & (k = 1, r+1), \\ (-)^k & (1 < k < r+1). \end{cases} \tag{10.43}$$

(10.37) follows from (10.39) and (10.43), and so the proof of the theorem is completed.

Note that we have at no point appealed to the fact that $T_r(u)$ has the specific form (10.30). Nevertheless, a good deal is now known about $T_r(u)$ (e.g. the number and nature of its extreme values). It is known that $u_1 = -1$, $u_{r+1} = 1$, while (10.36) and (10.37) between them determine all the other roots u_k, and so determine $T_r(u)$.

Theorem (10.7) is due to Kiefer and Wolfowitz (1959); the first author has also developed the idea considerably in later publications.

10.6 MATRIX GAMES AND THEIR EXTENSION: RELATION TO LINEAR PROGRAMMING

Consider a game (zero-sum, two-person) in which each player only has a finite number of pure strategies open to him. Such a game is termed a *matrix game*. Let the strategies open to players 1 and 2 be indexed by integers k and j respectively, and let a_{jk} denote the pay-off to player 1 if this pair of strategies is adopted.

Now, the function of two variables a_{jk} will not in general have a saddle-point, and the game in pure strategies thus need not have a value. However, consider the possibility that each player may randomize his strategy, so that player 1 chooses strategy k with probability x_k, and player 2, independently choosing from his set, assigns strategy j probability y_j. Then the game with the expected pay-off function

$$\phi(x, y) = \sum_j \sum_k a_{jk} y_j x_k = y^{\mathrm{T}} A x \tag{10.44}$$

certainly does have a value, because ϕ satisfies the conditions of the min-max theorem. Being bilinear, it has the required convexity and concavity properties: the vectors x and y are chosen from the sets $\mathscr{X}$ and $\mathscr{Y}$ of distributions over the pure strategies, and also have the properties demanded in Theorem (10.6). (That is, $\mathscr{X}$ is the set of points x for which $x \geqslant 0$, $\sum x_k = 1$; correspondingly for $\mathscr{Y}$.)

The statement that the game has a value is equivalent to the fact that the "order-of-play" is now irrelevant. That is, one player could announce his strategy (as the choice of a distribution over pure strategies) and the other could then react to his own best advantage, but the order in which the players might do this makes no difference to the average pay-off. In fact, no such announcement need be made—each player can work out for himself what strategies the other would find optimal, and the two optimizations can be regarded as simultaneous.

Because of the bilinear nature of ϕ, the conditions that $\bar{x}, \bar{y}$ be an equilibrium point take a specially simple form.

Theorem (10.8)

The necessary and sufficient conditions that $\bar{x}, \bar{y}$ be an equilibrium point of the game and v its value are that

$$\sum_j a_{jk} \bar{y}_j \leqslant v, \tag{10.45}$$

with equality if $\bar{x}_k > 0$, and

$$\sum_k a_{jk} \bar{x}_k \geqslant v, \tag{10.46}$$

with equality if $\bar{y}_j > 0$.

The conditions are plainly sufficient, in that they immediately imply $v = \phi(\bar{x}, \bar{y})$ and (10.14). To establish necessity, note that, by the definition of equilibrium, $\bar{x}$ maximizes $\phi(x, \bar{y})$ for x in $\mathscr{X}$. Allowing for the constraint $\sum x_k = 1$ by introduction of a Lagrangian multiplier λ, one sees that $\bar{x}$ maximizes the linear (in x) form $\phi(x, \bar{y}) - \lambda \sum x_k$ in $x \geqslant 0$. Consequently

$$\sum_j a_{jk} \bar{y}_j \leqslant \lambda, \tag{10.47}$$

with equality if $\bar{x}_k > 0$. It follows from this condition that $\lambda = \phi(\bar{x}, \bar{y}) = v$, so that condition (10.45) is necessary. Correspondingly for (10.46).

The characterization of an equilibrium point in Theorem (10.8) leads to a formulation of a matrix game as a linear programming problem, so reversing the interpretation of Section (10.4). By $\mathscr{X}$ and $\mathscr{Y}$ are understood the sets of respective distributions.

Theorem (10.9)

The value v of a matrix game can be characterized as the greatest value of λ which satisfies

$$\sum_k a_{jk} x_k \geqslant \lambda \tag{10.48}$$

for some x in $\mathscr{X}$, or as the least value of λ which satisfies

$$\sum_j a_{jk} y_j \leqslant \lambda \tag{10.49}$$

for some y in $\mathscr{Y}$. These two linear programming problems are mutually dual.

For, consider a j value for which $\bar{y}_j > 0$, so that equality holds in (10.46). Subtracting this equality from the inequality (10.48) at the same j value, one obtains

$$\lambda - v \leqslant \sum_k a_{jk}(x_k - \bar{x}_k) \tag{10.50}$$

so that

$$\bar{y}_j(\lambda - v) \leqslant \bar{y}_j \sum_k a_{jk}(x_k - \bar{x}_k) \tag{10.51}$$

for all j. Adding (10.51) over j, one deduces that

$$\lambda - v \leqslant \phi(x, \bar{y}) - \phi(\bar{x}, \bar{y}) \leqslant 0 \tag{10.52}$$

so that $\lambda \leqslant v$. Since equality ($\lambda = v$) is attainable, by Theorem (10.8), the first characterization of v then holds. The second follows analogously; verification of the final assertion is left to the reader.

One advantage of this formulation is the computational one: A game can be solved numerically by applying an algorithm such as the simplex method to one of the linear programming problems of Theorem (10.9).

Exercises

(10.19) Note that Exercise (10.8) solves the game of hide-and-seek in mixed strategies; it is known from Exercise (10.12) that a solution in pure strategies is impossible. Consider a version of this game in which the pay-off to the seeker if he finds the hider varies with the hiding-place.

(10.20) Solve the horse-backing/doctoring problem of Exercise (10.13) in mixed strategies. Interpret the solution.

(10.21) Exhibit the two LP problems of Theorem (10.9) as "complementary variational principles" deduced from the existence of a saddle-point of ϕ, as at the end of Section (10.4). (Actually, this is just a rewording of our general proof of the Lagrangian principle for the particular linear programme: maximize λ subject to (10.48) for distributions x.)

(10.22) The function $\phi(x, y) = y^{\mathrm{T}} A x - \rho y^{\mathrm{T}} x$ has a saddle-point, for any square matrix A and real scalar ρ, where x and y are distributions. Choose ρ so that the value of ϕ at this saddle-point is zero, and suppose that $y^{\mathrm{T}} x > 0$ at the saddle-point. Hence show that the function $\psi(x, y) = (y^{\mathrm{T}} A x)/(y^{\mathrm{T}} x)$ has a saddle-point with value ρ for x, y non-negative vectors such that $y^{\mathrm{T}} x > 0$, even though ψ does not satisfy the convexity-concavity assumption of Theorem (10.6). The importance of ρ has been noted in Section (5.1).

(10.23) *Proof of the min-max theorem for matrix games.* Let A be an $m \times n$ matrix, and $\mathscr{X}$ the set of vectors describing a distribution on n points: vectors $x \geqslant 0$ such that $\sum x_k = 1$. Suppose v is the largest value of λ consistent with (10.48), for some x in $\mathscr{X}$. This could be expressed by saying that the m-vector $(v, v, ..., v)$ is a boundary point of the convex set $A\mathscr{X} - R_+^m$. There thus exists a supporting hyperplane to the set at the point. That is, if the point is expressed $A\bar{x} - \bar{t}$ for $\bar{x}$ in $\mathscr{X}$ and $\bar{t}$ in R_+^m then there exists a vector $\bar{y}$ such that $\bar{y}^T(Ax - t)$ is maximal for (x, t) in $(\mathscr{X}, R_+^m)$ at $(\bar{x}, \bar{t})$. Hence show that the function $y^T Ax$ has a saddle-point $(\bar{x}, \bar{y})$, if x and y take values on the sets of distributions on n and m points respectively.

10.7   DECISION-MAKING UNDER UNCERTAINTY: ADMISSIBLE STRATEGIES AND BAYES STRATEGIES

The concepts of vector maximization and efficiency sketched in Section (10.1) find one of their principal applications in statistical decision theory. This theory concerns itself with the situation in which one of a number of actions has to be taken, the appropriateness of any particular action depending on certain unknown factors in the situation; the "state-of-nature". For example, if one is betting on a horse-race, the action is the determination of the amount of money to be staked on each horse. The appropriateness of any given bet will depend upon the relative strengths of the horses running, which one does not know, although experience may afford some guidance.

Suppose that there are N states-of-nature; let the supposition that the ith of these holds be termed the ith *hypothesis*, and be denoted H_i. Suppose that, if action x is taken and H_i holds, then there is an expected gain in utility of $f_i(x)$. For example, $f_i(x)$ might be one's expected winnings if one lodged a compound bet x when the true relative strengths of the horses, condition of the track, etc., were those asserted by H_i. (The term *expected* winnings is used, because the race has yet to be won, and one must average over the uncertainties of the race.)

In decision theory one tends to think in terms of losses rather than of gains, so the problem will be formulated as one of minimizing the losses

$$L_i(x) = -f_i(x) \tag{10.53}$$

instead of maximizing the utilities $f_i(x)$. The problem is one of vector minimization; one would wish to choose x so that $L_i(x)$ is small for all i. As in Section (10.1), such a simultaneous minimization is in general impossible, and so the notion of an efficient point must be regarded as the best compromise; those x must be considered for which there exists no x' in $\mathscr{X}$ such that

$$L(x') < L(x). \tag{10.54}$$

Here $L(x)$ is the vector with elements $L_i(x)$; the set $\mathscr{X}$ is the set of possible actions, or *strategies*, and in this context an efficient point x is termed an *admissible strategy*.

If the set $L(\mathscr{X}) + R_+^N$ is convex, then it is known from Theorem (10.1) that the admissible strategies are those which minimize a form

$$\pi^T L(x) = \sum \pi_i L_i(x) \tag{10.55}$$

in $\mathscr{X}$, for some $\pi > 0$. This has an interesting statistical interpretation. If the scale of π is standardized so that $\sum \pi_i = 1$, then π could be regarded as a probability distribution over states-of-nature, and expression (10.55) regarded as a loss averaged with respect to this distribution. That is, without having at all admitted the idea that i, the label of the "true" state-of-nature, could be regarded as a random variable, we have been led, via the concept of admissibility, to formally treat it as one. In minimizing expression (10.55), we are acting as though hypothesis H_i could be assigned a probability π_i, that the loss over the various contingencies is averaged on the basis of this distribution, and then a strategy chosen minimizing this average loss. It is characteristic of the *Bayesian* approach to decision-making that, from other considerations, i is frankly regarded as a random variable with some distribution $\{\pi_i\}$ (the *prior* distribution); it is interesting that the notion of admissibility leads to at least a formal adoption of such a viewpoint. The interpretation of π_i as a probability is reinforced in Exercise (10.25).

It is known (see Exercise (10.3)) that the admissible strategies are in fact just those which minimize expression (10.55) fors ome $\pi \geqslant 0$, or a limit of such a case. That is, they are the strategies or limit strategies derived on the supposition that all hypotheses have positive prior probability. The Bayes strategies x are those minimizing expression (10.55) for some $\pi > 0$, i.e. for any prior distribution, and are just those for which there is no x' in x such that

$$L(x') \leqslant L(x). \tag{10.56}$$

In other words, these are the strategies which are at least as good as any other under some circumstances (i.e. for some i).

No reason has yet been given for assuming $L(\mathscr{X}) + R_+^N$ convex. The justification is very much that which emerged in Section (3.8): if the set is not convex, then it is to be made so by the introduction of *mixed strategies*. Suppose, for example, that there are only finitely many basic actions which are possible, indexed by $k = 1, 2, \dots$. Suppose that L_{ik} is the expected loss under H_i if action k is taken. With this discrete set of actions, the loss set will certainly not be convex in general (see Exercise (10.24)). However, it is possible to *randomize*, and take action k with probability x_k (or a proportion

x_k of the time). The expected loss (or long-term average loss) is then

$$L_i(x) = \sum_k x_k L_{ik}$$

The vector $x = (x_1, x_2, \ldots)$ describes a distribution (and so takes values in a convex set $\mathscr{X}$). Since $L(x)$ is linear, $L(\mathscr{X})$ will then be convex. By extending the notion of a strategy, we have thus achieved a convex loss-set.

The strategy in which only a single one of the basic actions k is used will be termed a *pure strategy*. Now, for any given π, among the x which minimize expression (10.55) there is certainly a pure strategy (because any hyperplane supporting a convex set meets it in at least one extreme point: see Exercise (3.13)).

So, it is certain that, among the Bayes strategies for a given prior distribution π, there is at least one pure strategy. However, this is not so for the often used *minimax strategy*; the strategy x which minimizes the maximum loss $\max_i L_i(x)$ in the space of mixed strategies. This strategy is plainly admissible, and so to be found among the Bayes strategies. However, one sees from Figure (10.3) that the strategy need not be pure.

Exercises

(10.24) Suppose that x can take only finitely many values. Then show that $L(\mathscr{X}) + R_+^N$ is convex if and only if there is one strategy $\bar{x}$ uniformly superior to all others, in that

$$L(\bar{x}) \leqslant L(x) \quad (x \in \mathscr{X})$$

(10.25) The decision problem is sometimes presented in the apparently more general form that one must take an action, having observed the value of a random variable ω whose distribution depends upon the state-of-nature i. For example, in the betting case, one might have seen the same horses run in previous races. Suppose that ω has a probability density $\rho_i(\omega)$ relative to a measure μ if H_i holds. The optimal action x will depend upon the observation ω, and so is expressible $x = x(\omega)$. The vector loss will depend upon this function

$$L(x(.)) = \left(\int L_i(x(\omega)) \, \rho_i(\omega) \, \mu(d\omega) \right),$$

which must then be chosen to constitute an efficient strategy with this loss vector. Show that, provided the necessary convexity conditions are fulfilled, $x(\omega)$ is the value of x minimizing $\sum_i \pi_i \rho_i(\omega) L_i(x)$ for some $\pi > 0$. This is again consistent with a Bayesian interpretation in which π_i is the "prior probability" of H_i, because, in such a case, the "posterior probability" of H_i once ω has been observed is indeed proportional to $\pi_i \rho_i(\omega)$.

(10.26) Suppose that there are only two hypotheses, and two actions which amount essentially to acceptance of one or other of these hypotheses, so that $L_{11} < L_{12}$, $L_{22} < L_{21}$. Suppose that an observation ω has been taken. Show then that, in the notation of Exercise (10.25), action 1 should be taken if ω is such that

$$\frac{\rho_1(\omega)}{\rho_2(\omega)} > \frac{\pi_2}{\pi_1}\left(\frac{L_{21} - L_{22}}{L_{12} - L_{11}}\right)$$

(cf. Section (5.2)).

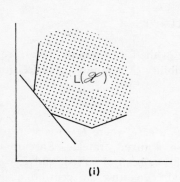

(i)

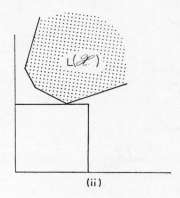

(ii)

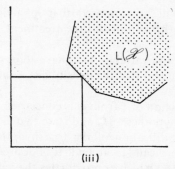

(iii)

Figure 10.3 A Bayes strategy is determined by the point L of $L(\mathscr{X})$ for which a linear form $\pi^T L$ is minimal. It is seen (case i) that there are always pure strategies achieving this minimum. For the minimax strategy, determined as the point where a hypercube erected on the co-ordinate axes first meets $L(\mathscr{X})$, this may or may not be true (see cases ii and iii)

(10.27) Show that to back all the horses in a race is never an admissible strategy.

(10.28) Suppose that a manufacturer receives a batch of components from a sub-contractor, and has two basic actions: to reject the batch or accept it. The expected losses consequent upon the two actions are a and $b\theta$ respectively, where a, b are constants, and θ the proportion of defective items in the batch. In order to derive some information on the value of the unknown θ, the manufacturer tests R items from the batch: the probability that he finds r defective is (conditional on θ) then known to be

$$\rho_r = \frac{R!}{r!\,(R-r)!}\theta^r(1-\theta)^{R-r}.$$

Show that an admissible strategy must be of the form: accept if $r < r_0$, reject if $r > r_0$, for some value r_0, with the possibility of a randomized decision if $r = r_0$.

Example

The following example is of interest, as a case where $L(\mathscr{X})$ is seen to be naturally convex, without appeal to the use of mixed strategies. Suppose that these are two random variables: ω, which has been observed, and ζ, which has yet to be observed. For example, ω might constitute the record of a signal up to a certain point in time, or the record of results of a sequence of learning experiments. The variable ζ would then constitute the future course of the signal, or of the sequence of experiments, respectively. It is desired to predict the value of ζ from that of ω.

For simplicity, it will be supposed that ω and ζ are scalars with a probability density in the ω/ζ plane: the treatment for more general cases follows by analogy.

Suppose that on H_i the variables ω, ζ have joint density $\rho_i(\omega, \zeta)$, that ω alone has density $\rho_i(\omega)$, and that the density of ζ conditional on the value of ω is $\rho_i(\zeta|\omega)$.

Now, as far as prediction of ζ from ω is concerned, one could do no better than identify the true hypothesis H_i, and state that ζ has conditional density $\rho_i(\zeta|\omega)$. However, identification of the true hypothesis H_i is not a prime objective, but merely a means to an end. It is desired to construct a probability density in ζ, $\nu(\zeta|\omega)$ which approximates the valid conditional expectation $\rho_i(\zeta|\omega)$ as well as possible in some sense.

Take the information measure

$$L_i(\nu) = -\iint \rho_i(\omega, \zeta) \log\left[\frac{\nu(\zeta|\omega)}{\rho_i(\zeta|\omega)}\right] d\omega\, d\zeta \qquad (10.57)$$

as a measure of the deviation between v and ρ_i: appeal to the continuous analogue of Exercise (3.21) shows that $L_i(v)$ attains its minimal value of zero if v equals ρ_i almost everywhere, $\rho_i(\omega, \zeta)$ measure.

The "action" in this case is the choice of a particular density v, from the set $\mathcal{N}$ of non-negative v satisfying

$$\int v(\zeta \,|\, \omega)\, d\zeta = 1 \tag{10.58}$$

for all ω. This set is convex, and, since $L_i(v)$ is a convex functional of v, the set $L(\mathcal{N}) + R_+^N$ is then also convex. Hence the conclusion of Theorem (10.1) can be invoked: any "efficient" or "admissible" v minimizes the average loss

$$\pi^T L(v) = -\int\!\!\int \sum \pi_i\, \rho_i(\omega,\, \zeta) \log\left[\frac{v(\zeta\,|\,\omega)}{\rho_i(\zeta\,|\,\omega)}\right] d\omega\, d\zeta \tag{10.59}$$

for some $\pi > 0$. Appealing again to Exercise (3.21), one sees that the v minimizing expression (10.59) must be given almost everywhere by

$$v(\zeta\,|\,\omega) = \frac{\sum_i \pi_i\, \rho_i(\omega,\, \zeta)}{\sum_i \pi_i\, \rho_i(\omega)}. \tag{10.60}$$

If π is assumed standardized so that $\sum \pi_i = 1$, then expression (10.60) has an immediate interpretation: the density of ζ conditional on ω if i is also regarded as a random variable with distribution $\{\pi_i\}$. In other words, any admissible choice of $v(\zeta\,|\,\omega)$ can be regarded as a density of ζ conditional on ω for some prior distribution over hypotheses.

References

Apostol, T. M. (1957). *Mathematical Analysis*. Addison-Wesley.

Arthurs, A. M. and Robinson, P.D. (1968). "Complementary variational principles for a generalised diffusion equation." *Proc. Roy. Soc.*, A, **303**, 497–502.

Beale, E. M. L. (1968). *Mathematical Programming in Practice*. Pitman.

Bellman, R. (1960). *Introduction to Matrix Analysis*. McGraw-Hill.

Berge, C. and Ghouila-Houri, A. (1965). *Programming, Games and Transportation Networks*. Methuen. Publication of French original in 1962, by Dunod.

Burger, E. (1959). *Introduction to the Theory of Games*. English translation, Prentice-Hall, 1963.

Chan, H. S. Y. (1968). "Minimum volume design of frameworks and discs for alternative loading systems." *Quarterly Appl. Math.*, **25**, 470–473.

Chernoff, H. and Moses, L. E. (1959). *Elementary decision theory*. Wiley.

Duffin, R. J., Peterson, E. L. and Zener, C. (1967). *Geometric Programming*. Wiley.

Ferguson, T. S. (1967). *Mathematical Statistics*. Academic Press.

Fiacco, A. V. and McCormick, G. P. (1968). *Nonlinear Programming: Sequential Unconstrained Minimisation Techniques*. Wiley.

Gale, D. (1968). "A Mathematical theory of optimal economic development." *Bull. Amer. Math. Soc.*, **74**, 207–223.

Isii, K. (1964). "Inequalities of the types of Chebyshev and Cramer–Rao and mathematical programming." *Ann. Inst. Statist. Math.*, **16**, 277–293.

Lefschetz, S. (1949). *Introduction to topology*. Princeton University Press.

Lindgren, B. W. (1960). *Statistical Theory*. Macmillan.

Karlin, S. (1959). *Mathematical Methods and Theory in Games, Programming and Economics*. Addison–Wesley, Pergamon.

Karlin, S. and Studden, W. J. (1966). *Tchebycheff Systems: with Applications in Analysis and Statistics*. Interscience Publishers.

Kiefer, J. (1953). "Sequential minimax search for a maximum." *Proc. Ann. Math. Soc.*, **4**, 502–506.

Kiefer, J. and Wolfowitz, J. (1959). "Optimum designs in regression problems." *Ann. Math. Stat.*, **30**, 271–294.

Künzi, H. P., Tzschach, H. G. and Zehnder, C. A. (1968). *Numerical Methods of Mathematical Optimisation*. Academic Press.

Michell, A. G. M. (1904). The limits of economy of material in frame-structures. *Phil. Mag.* (6), **8**, 589–597.

Noble, B. (1964). *Univ. Wisconsin Math. Res. Center Rep.* no. 473.

Roode, J. D. (1968). *Generalised Lagrangian functions in Mathematical Programming*. "Bronder-Offset", Rotterdam.

Valentine, F. A. (1964). *Convex Sets*. McGraw-Hill.

Von Neumann, J. and Morgenstern, O. (1944). *Theory of Games and Economic Behaviour*. Princeton U.P.

Žáčková, J. (1966). "On minimax solutions of stochastic linear programming problems." *Časopis pro pěstování matematiky*, **91**, 423–430.

Index

Page numbers given in italics indicate the main discussions of an entry